The Dialectical Primatologist

D1481444

The Dialectical Primatologist identifies the essential parameters vital for the continued coexistence of hominoids (apes and humans), synthesising primate research and conservation in order to develop culturally compelling conservation strategies required for the facilitation of hominoid coexistence.

As unsustainable human activities threaten many primate species with extinction, effective conservation strategies for endangered primates will depend upon our understanding of behavioural response to human–modified habitats. This is especially true for the apes, who are arguably our most powerful connection to the natural world. Recognising the inseparability of the natural and the social, the dialectical approach in this book highlights the heterogeneity and complexity of ecological relationships. Malone stresses that ape conservation requires a synthesis of nature and culture that recognises their inseparability in ecological relationships that are both biophysically and socially formed, and seeks to identify the pathways that lead to either hominoid coexistence or, alternatively, extinction.

This book will be of keen interest to academics in biological anthropology, primatology, environmental anthropology, conservation and human–animal studies.

Nicholas Malone is a senior lecturer in Anthropology at the University of Auckland, New Zealand. Dr Malone is an anthropologist with a broad interest in the social and ecological lives of primates, especially those of apes and humans. Specifically, he seeks to understand how the observed patterns of variability within and between taxa are simultaneously shaped by, and act as shaping factors of, evolutionary processes. Additionally, he strives to contribute to primate conservation through a commitment to engaging with local and extra-local efforts. Finally, he wishes to situate the study of primates within the broader contexts of anthropology, history and research ethics. His writing is informed by research experiences in Indonesia and the Democratic Republic of the Congo.

New Biological Anthropology

Series Editor: Agustín Fuentes

University of Notre Dame, USA

Emergent Warfare in Our Evolutionary Past
Nam C. Kim and Marc Kissel

The Promise of Contemporary Primatology
Erin P. Riley

Genetic Ancestry
Our Stories, Our Pasts
Jada Benn Torres and Gabriel A. Torres Colón

The Dialectical Primatologist
The Past, Present and Future of Life in the Hominoid Niche
Nicholas Malone

For more information about this series, please visit: www.routledge.com/New-Biological-Anthropology/book-series/NBA

The Dialectical Primatologist

The Past, Present and Future of Life in the Hominoid Niche

Nicholas Malone

Routledge
Taylor & Francis Group

LONDON AND NEW YORK

First published 2022
by Routledge
2 Park Square, Milton Park, Abingdon, Oxon OX14 4RN

and by Routledge
605 Third Avenue, New York, NY 10158

Routledge is an imprint of the Taylor & Francis Group, an informa business

© 2022 Nicholas Malone

The right of Nicholas Malone to be identified as author of this work has been
asserted by him in accordance with sections 77 and 78 of the Copyright,
Designs and Patents Act 1988.

All rights reserved. No part of this book may be reprinted or reproduced or utilised
in any form or by any electronic, mechanical, or other means, now known or
hereafter invented, including photocopying and recording, or in any information
storage or retrieval system, without permission in writing from the publishers.

Trademark notice: Product or corporate names may be trademarks or registered trademarks,
and are used only for identification and explanation without intent to infringe.

British Library Cataloguing-in-Publication Data
A catalogue record for this book is available from the British Library

Library of Congress Cataloging-in-Publication Data
A catalog record has been requested for this book

ISBN: 978-0-367-21124-0 (hbk)
ISBN: 978-0-367-21132-5 (pbk)
ISBN: 978-0-367-21134-9 (ebk)

DOI: 10.4324/9780367211349

Typeset in Bembo
by Newgen Publishing UK

For Margaux.

Contents

List of illustrations   viii
Foreword   xi
Acknowledgements   xiv

1 Introduction: The dialectical primatologist   1

2 From the Miocene to the margins: Overview of the
   superfamily Hominoidea   26

3 Emergence: Theorising ape sociality   64

4 Waves of change: Insights from Java, Indonesia   96

5 Betwixt and between: Apes in (and on) the verge   129

6 Conclusion: The future of life in the hominoid niche   160

   Epilogue: The view from Aotearoa New Zealand   181

   Index   187

Illustrations

Figures

1.1 Infant agile gibbon (*Hylobates agilis*) as presented for sale in the Pramuka Market, Jakarta, July 2000   10

1.2 Flow chart of required post-arrival procedures, subsequent to receiving approval of the research permit   11

2.1 Analysis of the past 20 million years reveals a pattern of relative species diversity of hominoids and cercopithecoids: monkey diversity has increased whereas hominoid diversity has declined   28

2.2 Fossil ape community reconstruction from Rusinga Island, Kenya, depicting multiple taxa from the family Proconsulidae   30

2.3 A comparison of the cranial remains of *Sivapithecus indicus* (GSP 15000) with the crania of both *Pan* (left) and of *Pongo* (right)   31

2.4 The Javan silvery gibbon (*Hylobates moloch*), just one of 20 recognised species of hylobatid   36

2.5 A wild (non-rehabilitated) adult female Sumatran orangutan (*Pongo abelii*) and offspring at Bukit Lawang   42

2.6 Mountain gorilla (*Gorilla beringei beringei*) in Rwanda   43

2.7 Two species within the genus *Pan*. Above: an adult male bonobo (*Pan paniscus*) "Larry" at the Iyema study site, DRC. Below: an adult female chimpanzee (*Pan troglodytes*) "Ekisigi" with her dependent offspring "Euro" in Kibale National Park, Uganda   45

2.8 The author and his daughter negotiating symbolic, somatic and material terrain in Yogyakarta, Indonesia   48

3.1 Axes of paradigmatic assumptions common in primatological research   67

3.2 Simple trajectory for expanding social plasticity. Specific aspects of cognition, especially those related to behavioural plasticity, evolved largely in response to the increasingly challenging demands of a social life that includes the construction, alteration and inheritance of social and ecological niches   74

3.3 A model for socioecological change through time that combines the effects of natural selection, niche construction and the locally specific, salient patterns of anthropogenic influences. Niche construction theory conceptualises evolution of *both* populations of organisms (O) *and* their environments (E) through reciprocal, causal bouts of natural selection (NS) and niche construction (NC) over generational time   77

4.1 Kang Heri's 1960s-era Land Rover on one of many trips over the precarious slopes of Gunung Gelap (which translates to "dark mountain") between Bandung and Pameungpeuk, circa 2005. Within moments, a sufficient crowd had gathered to assist in righting the ship. With a turn of the engine crank we were back on our way   97

4.2 Banteng (*Bos javanicus*) at the Cidaon grazing area, Ujung Kulon National Park   105

4.3 Sacred sites, or *keramat*, within the Leuweung Sancang. On the left is the Cikajayaan waterfall, easily the most popular pilgrimage location within the forest. Pilgrimage visits vary in length from several hours to several months, with stated goals ranging in scope from simply bathing in the sacred waters, to attempts to "tilem – menghilang dari dunia" (disappear from earth, or become enmeshed with the spiritual world). On the right is an adult male silvery gibbon, as photographed from a less-frequently visited sacred site   107

4.4 The location of the Cagar Alam Leuweung Sancang (CALS) in West Java, and a detail of the major geographical features within the reserve boundary   108

4.5 The effects of tsunami waves on the night of 22 December 2018 on both the peninsula and mainland portions of Ujung Kulon National Park. The photo on the left, near Citelang, reveals the extent of destruction to coastal trees (extending approximately 150 metres inland from the inter-tidal zone). The photo on the right is of government-provided housing, near Tamanjaya, for coastal-wary and/or displaced communities. With only short distances between the sea and the protected reserve boundary, a difficult balance is being struck between safety, livelihoods and ecosystem impacts   111

4.6 The present boundaries and major regions of Ujung Kulon National Park, including (from left to right): Panaitan Island, the peninsula and the mainland (including the Honje ranges). The most prominent areas on the map indicate the Jungle (*light grey*), Core (*medium grey*) and Marine Protected (*dark grey*) Zones   113

4.7 The pilgrimage site of Cimahi in the Legon Pakis management area of Ujung Kulon National Park. Pictured at right is the grave of a saint (one of five) in this *makam keramat* (sacred burial ground)   115

4.8   Reconnaissance walks along the forest–field transitional
      boundaries to map the extent of agricultural encroachment
      into the National Park   117
5.1   Left: a rehabilitated and reintroduced male silvery gibbon
      (*Cheri*) accessing a temporary feeding station at a release site.
      Right: the wild female (*Dewi*) also learned to access the feeding
      station. After his release partner (*Ukong*) required evacuation
      and a return to the rehabilitation centre, *Cheri* and *Dewi*
      became a stable pair   134
5.2   Then graduate student (now PhD, Auckland 2020), Alison
      Wade during her research to assess local attitudes towards
      conservation initiatives in West Java. Alison is accompanied by
      long-term field assistant Ajat Surtaja (right)   137

Tables

2.1   Taxonomy of the living members of the superfamily
      Hominoidea   34
2.2   Taxonomy of the family Hylobatidae   35
2.3   Taxonomy of the family Hominidae   40
3.1   Comparison of parameters and predictions between SET
      and EES   71
4.1   Comparison of population parameters for silvery gibbons
      in the Cagar Alam Leuweung Sancang (CALS) and Ujung
      Kulon National Park (UKNP). At both sites, higher group and
      individual densities are associated with the common presence
      of spiritual pilgrims   115

Foreword

to Nicholas Malone, *The dialectical primatologist: The past, present and future of life in the hominoid niche*

Redefining humans as animals with a direct kinship to and ecological entanglement with other primates and animals, as Nicholas Malone sets out to do, is a recent development, not just in the history of primate studies but also in western consciousness at large. The latter has long struggled, and arguably still is struggling, to rid itself of deeply engrained, Eurocentric and anthropocentric notions of human specialty and a rigorous nature-culture divide.

When in the 18th century Linnaeus classified humans with newly discovered great apes this was vehemently criticised and quickly reversed. A hundred years later Darwin's idea of apish ancestry was an equally controversial redefinition of the animal–human boundary, and neutralised by excluding reason and moral feelings, or the human soul, from the process as unique human features.

The rebuff of alloprimates continued in 20th-century anthropology, which has been described as the most humanistic of the sciences and the most scientific of the humanities. Anthropology keeps struggling for intellectual and disciplinary coherence in its efforts to negotiate deep tensions between interpretive and explanatory approaches and, in fact, between two age-old metaphysical views of human nature that feed into these methodological stances: one stressing its continuity and one stressing its unique rupture with nature.

For me as a philosopher with a longstanding interest in primate studies and their checkered history this essay by a primatologist with a longstanding interest in philosophy was an exciting read. From my perspective, ethnoprimatology is the most recent negotiation of the sacrosanct human–animal boundary. This emergent paradigm, like convergent initiatives from the last few decades such as Multispecies Ethnography, Animal Studies and One Health, constitutes a new arena for efforts to bridge disciplinary and sub-disciplinary rifts between and in anthropology's four fields.

How much the aforementioned tensions in anthropology continue is clear from the fact that both factions often, if not usually, do not see eye to eye. Each has its own methodologies, conferences, journals and conception of disciplinary identity. Hostility and redlining ("postmodernists!"; "reductionists!") abound in most anthropology departments worldwide, despite sustained efforts to avoid too one-sided approaches, for example in British social anthropology or in fieldwork-based ethnographies. In the United States, there have been several departmental splits in anthropology along these lines since the 1990s.

Interestingly, ethnoprimatology draws on both sides of the epistemic divide. On the one hand on Boasian and (neo-)Durkheimian cultural anthropology, which likes to think of itself as a human (and humane) science, in a non-, if not anti-Darwinian spirit. When studying its development during the 20th century, it is hard to avoid the impression that the repudiation of life sciences approaches at least partly has been, and still is, constitutive of cultural anthropology's disciplinary identity. On the other hand, ethnoprimatology has recourse to behavioural socioecology and biological/evolutionary anthropology, including primatology, long under the sway of Sherwood Washburn's New Physical Anthropology, associated with the mid-20th-century New Synthesis in the life sciences.

Nicholas Malone's anti-reductionist, holistic and historicist ethnography of human–alloprimate entanglements and his project of reclaiming primatology as anthropology seem to situate him with the (culturalist) cultural anthropologists – faced with a major, seemingly unsurmountable challenge to work out how this position combines with life sciences approaches.

However, here's the good news, which is threefold. In the first place, Malone is epistemologically sophisticated in not just, as a practicing primatologist, being aware of fundamental conceptual and methodological issues, but explicitly addressing the latter in the present essay. The term "epistemology", sometimes used in a rather vague manner, refers to the analysis not of data directly but of how it is handled conceptually, in terms of fundamental presuppositions of researchers. Therefore, his essay at least has the merit of explicitly engaging with the aforementioned epistemic divide in anthropology head-on, whatever problems may remain.

Secondly, the fact that he, as far as the life sciences are concerned, explicitly relies no longer on the classic, mid-20th-century New Synthesis (of the theory of evolution and genetics), but on the Extended New Synthesis, is a smart move which holds great promise. This new paradigm seeks to soften the New Synthesis' reductionist and nomothetic stress on bottom-up genetic causality by focussing on epigenetic processes, including cultural niche construction, with much room for what is historically emergent and contingent. This brings the life sciences closer than ever to ethnographic analysis.

A third promising aspect is Malone's dialectical approach, which is rooted in a long philosophical tradition reaching back to Greek thought: his way of engaging with, and connecting, deeply embedded and seemingly opposite dichotomous categories like natural/social; ethnography/primatology; wild/captive; genetic/emergent; as well as organisms as both subject and object, both cause and effect of their environment.

For every scientific discipline it makes sense every once in a while to take a step back to reflect on its current state and future direction. *The dialectical primatologist* constitutes a major self-reflexive moment. The book's sustained attention to ethical issues – stressing compassionate coexistence in shared, multispecies niches – adds to its importance in the Anthropocene, an era of unprecedented transformation of an entire planet by a single primate species to suit its own needs.

The dialectical primatologist, finally, is autobiographical not just as serious academic self-reflection but also as a delightful ethnographic travelogue in which this particular reader relived much from his own explorations in Java, Sumatra and Bali.

Raymond Corbey
(Professor of Philosophy of Science and Anthropology,
Leiden University, the Netherlands)

Acknowledgements

West Java is my home away from home. Arguably, a foreign researcher's most valuable asset is a network of trusted local counterparts and skilled research assistants. In Bandung, I owe a massive debt of gratitude for the patience and support of so many. In this regard, I wish to acknowledge Made Wedana, Asep Purnama, Hasadungan Pakpahan, Mang Ajat (Ayut) Surtaja, Heri Oktavinalis, Nissa Nuraini, Kang Heri UU, Kang Ipan Juanda, Ery Bukhorie, Sapari Varlin, Yana, Danius, Djuniwarti, Priet, Mely, Indah Winarti and Sigit Ibrahim – *hatur nuhun pisan*. All of you have opened up your world (and your homes) to myself, my family and my students. For that, and many more small favours, I am forever grateful. I wish to thank my extended family in the Sancang Village (including three generations of the Momon Haes lineage). Alongside you I have spent many years obtaining insights about animals, spirits, trees and tempeh. Your friendship, hospitality, logistical support and local knowledge are very much appreciated. Most recently, my research in Banten Province was only possible because of the collaboration with staff at Universitas Padjadjaran, especially Professor Johan Iskandar, Dr Teguh Husodo, Dr Eneng Nunuz Rohmatullayaly and the late Dr Ruhyat Partasasmita. Within Ujung Kulon National Park, I am thankful for the research assistance of Sidik Permana and the local logistical support of Pak Nana, Pak Satun and Pak Encud.

Born and raised in Chicago, I acknowledge my working class, Midwest upbringing for providing a grounded attitude towards work and inter-personal relationships. I graciously thank my late mother Sandi, my father Drew, step-mother Pam, brother Mark and all my extended family for your constant support. Outside of family and school life, I pursued outdoor activities at the local "mountain" (actually an ancient glacial moraine) on the Illinois/Wisconsin border and developed lifetime friendships. Although our adventures are now fewer and farther between, I am grateful to friends Brad and Samantha Thein, John and Heather Carlson, Kevin and Jackie Brady and Mike Fleming for the decades of lasting memories and shared experiences.

Eventually on to Colorado (and actual mountains) at age 18, my horizons were expanded by the likes of Kirk Reichard, René Crout, Bobby Walker and Theo Bessin and the late Jonny Copp. At the University of Colorado, I studied anthropology and developed an interest in primatology thanks to Michelle

Sauther's courses on primate behaviour and ecology. Her fascinating stories of lemurs and Madagascar's forests piqued my interest and imagination, and her insightful lessons on the ecological underpinnings of primate social systems motivated my early studies. For my MSc I moved to the state of Washington, and at Central Washington University I was fortunate to find a lifelong mentor and friend in Agustín Fuentes. From Ellensburg to Bali, and from South Bend to Auckland, I owe much of my teaching and research acumen to your guidance and support. Almost 20 years ago you introduced me to the incredibly complex endeavour of conducting anthropological research in Indonesia, and I have been returning regularly ever since. Additionally, in Washington I met a dedicated cadre of animal rights activists and friends. To John Mulcahy, Diana Goodrich and Sarah Baeckler, I am constantly impressed and inspired by your dedication to improving the lives of animals, especially in securing dignified lives for chimpanzees freed from the confines of biomedical research.

I pursued my doctorate at the University of Oregon in Eugene. My experience in Oregon was greatly enriched by my supervisors, teachers and a cohort of fellow graduate students and friends. First and foremost, thank you to Frances White for your expert guidance and for the opportunities you have afforded me. Thanks to John Bellamy Foster, Stephen Frost, Joanna Lambert, John Lukacs, Josh Snodgrass, Stephen Wooten and Richard York for additional guidance and inspiration. Thank you to Melissa Baird, Tara Cepon-Robbins, Jen Erickson, Patrick Hayden, Stefano Longo, Felicia Madimenos, Shayna Rohwer, Britta Torgrimson, Carolyn Travers and Michel Waller – we sure lived and learned, didn't we?

Since 2010, I have found a home in the anthropology department at the University of Auckland in Tamaki Mākarau, Aotearoa/New Zealand. My heartfelt appreciation to Judith Littleton, Bruce Floyd and Heather Battles – your collective insight and support have been invaluable. Although we take our work seriously, we do not take ourselves too seriously. And for a profession prone to stress, isolation and self-doubt, you have always made coming to work a fun and functional proposition. As newly arrived immigrants to Aotearoa, Julie Park made Meredith and I feel welcome and part of the *whānau*. Collectively, Harry Allen, Melinda Allen, Ethan Cochrane, Simon Holdaway, Judith Huntsman, Sunhee Koo, Thegn Ladefoged, Rebecca Phillips and Peter Sheppard have all contributed to the happiness we have found in the land of the long white cloud – *ngā mihi nui*. At the University of Auckland I have been fortunate to supervise a superb and hard-working group of postgraduate research students. To my first doctoral descendant Alison Wade, as well as Megan Selby and masters-level students Ally Palmer, Courtney Addison, Nipuni Wijiwardene, Laura Taylor, Kathryn Ovenden, Jen Hale, Lotte van den Hout and Holly Steiner – your curiosity, talents and interests have pushed me to become a better academic, and I am proud of all your respective accomplishments. A special thanks goes to Amy Robbins and Christine Tintinger at the Auckland Zoo for your support of my research students both in Auckland and in Sumatra. Cheers to the biological anthropology writing group for your useful feedback and productive time spent at Old Government House, and to Pauline Herbst for introducing me to Scrivener writing software. In the latest stages of the book's development, Dr Alison Wade provided expert advice as well as editing and indexing services. Finally, at Auckland we are fortunate to have the support of our technical staff in anthropology. Thanks to Dr Natalie Remedios, Dr Rod Wallace, the late Vernon Tintinger, Tim Mackrell, and our illustrators, Seline McNamee and the late Briar Sefton.

A scholar's research trajectory is influenced in a myriad of ways, both large and small, through intellectual exchange, conversations and collaborations. Within anthropology and primatology, I extend my gratitude to (in alphabetical order): Thad Bartlett, Simon Bearder, Alison Behie, Richard Bergl, Aletta Biersack, Susan Blum, Catherine Bolten, Klaree Boose, Colin Brand, Mark Busse, Christina Campbell, Melissa Cheyney, Zana Clay, Amy Cobden, Raymond Corbey, Sienna Craig, Sharyn Graham Davies, Kerry Dore, Jef Dupain, Christine Dureau, Wendy Erb, Joseph Fracchia, Roger Fouts, Lee Gettler, Donna Glowacki, the late Colin Groves, Cyril Grueter, Darcy Hannibal, Alexana Hickmott, Kimberley Hockings, Ali Hofner, Michaela Howells, Carolyn Jost Robinson, Deborah Judge, Eben Kirksey, Kevin Langergraber, Susan Lappan, Lydia Light, James Loudon, Katie MacKinnon, James McKenna, David McMurray, Justyna Miszkiewicz, Geraldine Moreno Black, Madonna Moss, Marama Muru-Lanning, Anna Nekaris, Vincent Nijman, Rahul Oka, Gerard O'Regan, Melissa Panger, Varsha Pilbrow, Michael Plavcan, Ulrich Reichard, Melissa Reisland, Melissa Remis, Erin Riley, Deborah Rotman, Noel Rowe, Julienne Rutherford, Crickette Sanz, Mark Schurr, Christopher Shaffer, Susan Sheridan, Jaima Smith, Karen Strier, Vania Smith-Oka, Lynn Stephen, Lawrence Sugiyama, Julie Teichroeb, Zaneta Thayer, Brian Tilt and Russ Tuttle. Each of the aforementioned scholars have contributed to my intellectual development, and I greatly appreciate our professional engagement and exchanges.

Over the years I have received generous funding for my research from the Endangered Species Fund of the Chicago Zoological Society, Sigma Xi, the American Society of Primatologists, the Center for Asian and Pacific Studies, as well as doctoral and postdoctoral fellowships from the University of Oregon, and Faculty Research and Development Funds from the University of Auckland. A special thanks to Sir Michael Fay and Auckland University's archaeological field school staff for extending your generous hospitality to me during my writing retreat to Ahuahu Great Mercury Island. Finally, I would like to thank the editorial team at Routledge for the opportunity to contribute to the New Biological Anthropology book series. I extend my utmost gratitude to Agustín for your encouragement and skilful guidance, to anonymous reviewers of the prospectus and the manuscript, and especially to Katherine Ong, Amy Doffegnies and Alexandra McGregor. Your collective efforts to see this work through to fruition are extremely appreciated. It is not a perfect book, and I take responsibility for any and all of the book's remaining errors or shortcomings.

Last but certainly not least, it was back in Boulder, in the mid-1990s, where I met Meredith. As a fellow anthropology major and core snowboarder, we had much to discuss and discover. Meredith became my lifelong partner, and together we co-parent the force of nature that is our beloved daughter Margaux. It is now through Margaux's eyes that we experience an enriched present and envision possible futures. For Margaux and her generation, I am determined to work towards a future that includes some semblance of the planet that I have been so fortunate to experience – complete with a resplendent array of plant and animal life.

1 Introduction

The dialectical primatologist

Entrance into the sacred heart of the Sancang forest requires a steep descent of 337 unevenly aligned, concrete steps. With every passing year, the tentacles of tree roots make further advances in their inevitable quest to reclaim the forest floor. *Ka handap; ka luhur* – the Sundanese (West Javan) terms for descending and ascending – I climb the steps several times a day following gibbons from pre-dawn to dusk. Branching-off from the steps are small trails that circumvent the buttresses of giant trees. The equatorial sun is filtered prism-like through a dense canopy that is characteristic of lowland tropical rainforests. This remnant pocket of intact forest is both a pleasing glimpse, and a painful reminder, of what was once widespread. For the gibbons overhead, this particular combination of dipterocarp (Dipterocarpaceae), dracantomelon (Anacardiaceae) and fig (Moraceae) species represents the ideal habitat for sleeping, singing, feeding and socialising. Footsteps and voices can be heard at the top of the steps – another group of *penziarah* (spiritual pilgrims) are entering the forest. This time it is a group of six men. Earlier in the day it was an extended family. Trailing the men I see Pak Ade, a prominent *kuncen* (spiritual intermediary, literally "keeper of the keys", or "gatekeeper"), who also happens to be the father of the village head. Without emotion or reaction, I draw a line through the last several minutes of recorded data and close my notebook. In a few moments, I will be compelled to greet the visitors, and discuss my recent experiences in the forest with Pak Ade. The ensuing disruption to the data collection regime will invalidate the present sample of meticulously recorded gibbon behaviour.

Sundanese society follows Javanese in that almost nothing is more important than the knowledge of, and adherence to, behavioural etiquette and expectations. After a short conversation while leaning against a formidable tree trunk, Pak Ade shifts his weight and strikes the tree with his *golok* (machete) to remove a strip of *kulit kayu palahlar* (bark from this particular dipterocarp species). Most assuredly, when dusk falls in the forest, the bark will be set alight and transformed into aromatic wisps of smoke to complement Pak Ade's ritual incantations. The perception of this incense will be short-lived, eventually succumbing to the diluting clouds of *cengkeh*-infused smoke from the non-stop crackling of kretek cigarettes – a ubiquitous reminder of the importance of clove (*Syzygium*

DOI: 10.4324/9780367211349-1

aromaticum) as a defining commodity of the Asian spice trade. Chants turn into sunset Maghrib prayers; prayers into story-telling and informal conversation. Twenty-five metres above, a group of four gibbons are settled-in for the night. Fatigued and lonely, though never alone, I copy my data by hand into back-up notebooks and struggle in vain to keep my equipment dry during the seasonal rains. After a few hours of sharing stories and sipping coffee, I excuse myself from the social moment and retire to my tent, all the while attempting to regulate my internal and external states.[1]

Years later, as a postdoctoral research fellow, I joined an expedition to the remote interior of Équateur (now Tshuapa) Province in the Democratic Republic of the Congo. Our team of researchers, led by Professor Frances White, a pioneer in the study of wild bonobos (*Pan paniscus*), began the difficult process of re-establishing research protocols that were interrupted by the devastating Congolese Wars of the late-20th and early-21st centuries (1996–2003). Through the counting of bonobo nests and the recording of direct contacts with these elusive apes, we pieced together an understanding of the post-war, bonobo population status. However, the perceived costs and benefits of obtaining these research insights, from this and other bonobo research sites, vary among the concerned parties: namely ourselves, the communities supporting our research and local administrative powers. These subjectivities were thrown into sharp relief when it became clear that forests such as this one, with a history of researcher presence, can become magnets for militants in search of resources. And during times of political upheaval and violence, this unwanted attention exposes those who remain present, local peoples and ape study-subjects alike, to potentially fatal consequences. Would our presence, our data and our conservation recommendations produce positive outcomes? Or would our activities generate a disruptive wake, subtly (or not-so-subtly) experienced well beyond our departure? In this high-stakes environment, long-term commitments are crucial.[2]

The above stories serve to demonstrate that primate research and conservation activities do not take place in isolation, and attempting to tease apart the ecological and social lives of primates, amidst the context of humankind's planet-altering reach, is exceedingly difficult work. My own career as a primatologist is based on a series of research engagements with captive, displaced and wild apes. From captive gorillas, chimpanzees and orangutans to wild populations of gibbons and bonobos, as well as displaced and rehabilitated apes within various stages of an illegal trade, I have for the past 20 years sought a more complete understanding of humankind's relationships with the apes. In doing this, a host of important questions arise. Is it right to habituate wild apes for scientific research purposes? Is it ethical to keep apes in captivity? Should we prioritise the health and welfare of ape populations over those of marginalised human communities? In both theory and method, divisions between scientific and humanistic epistemologies are porous; and independently they are untenable as discreet modes of primatological enquiry. Finding answers to these complex questions will require the full breadth of the anthropological toolkit.

Few, if any, primatologists enter the field adequately equipped for what lies ahead. In this regard, I will argue that primatology emerging from the discipline of anthropology, as opposed to other common, parent disciplines, affords some distinct insights and advantages.[3] Appropriately, the *raison d'être* for homing the study of nonhuman primates within a discipline concerned with humanity is the premise that there are important comparative insights to be obtained. This is especially salient when our desire to know ourselves and our place in nature is powerfully mediated through our interactions with apes. The apes (co-members alongside humans in the superfamily Hominoidea) are arguably our most powerful connection to the natural and social worlds we occupy (Haraway 1989; Corbey & Theunissen 1995; Corbey 2005; Fuentes 2012). In the concluding pages of *The Metaphysics of Apes*, Raymond Corbey (2005: 200–201) expounds on the potential of apes to serve as:

> [m]issing links, in a new, more positive key: no longer as links in the Great Chain of Being, as was still the case in Linnaeus' natural history, more primarily as evolutionary links in an "ascent" to civilisation, but as go-betweens and mediators between humans and other animals, philosophically, scientifically, and morally. Like the trickster figures from the myths and rituals of so many peoples worldwide – Maui among the Polynesians, Semar among the Javanese, the Raven among American Indians and Arctic peoples – they can break taboos, transgress borders, and bring together opposites as benefactors.

As a clear starting point, I follow Marks (2011) in perceiving the study of apes to be a deeply cultural endeavour. From colonial imaginations to folklore and fiction, from anatomical and genetic comparisons to taboos and talismans, our inter-relationships with apes are complex and varied.

Dialectics and the philosophy of science

In primatology, tensions among preconceptions, academic preparedness and real-world experiences are heightened through the often transformative processes of fieldwork (MacClancy & Fuentes 2011). As an anthropological primatologist, I have welcomed the moments of cultural- and species-dissonance when they are arise, and embraced the resulting "habits of thought" that emerge from the awareness of both inherent complexities and ever-present contradictions within our various modes of operation. Similarly, the biologists Levins and Lewontin (1985: 267) attempt to explicitly state their intellectual preconceptions by describing the tension between "the materialist dialectics of our conscious commitment and the mechanistic, reductionist, and positivist ideology that dominated our academic education and that pervades our intellectual environment". In their research and writing, these preconceptions and tensions emerge as certain "habits of thought, certain forms of questioning that we identify as dialectical" (1985: 267). I suspect these habits of thought, an internal dialectic, are commonplace among primate researchers and conservationists. The self-description for the foundations of this book is that of the *dialectical primatologist*, and my positions are crafted by theoretical and practical contributions from a holistic anthropological tradition, with value added from historical, philosophical and sociological perspectives. Therefore, I explicitly ground this book in the traditions of an anthropological primatology and seek to extend their impact. Indeed, such an approach finds resonance with several themes so adeptly articulated in Erin P. Riley's (2020) *The Promise of Contemporary Primatology*, a preceding book in the New Biological Anthropology (Routledge) series.

The parallels with previous contributions to theorising evolution and the social dimensions of science are unmissable, specifically *The Dialectical Biologist* (Levins & Lewontin 1985) and *Biology Under the Influence* (Lewontin & Levins 2007). Heavily influenced by their colleague at Harvard, Stephen Jay Gould, and his prodigious work (e.g., 1977, 1989, 2002), these scholars sought to problematise a narrowly formed, deterministic and reductive view of biology by advocating for explicit consideration of historical, emergent and contingent processes.[4] A dialectical approach does not preclude reductionism, strategically deployed, which is indeed an efficacious approach to knowledge acquisition. However, in embracing a dialectical approach, one must abandon reductionism as an exclusive mode of operation. I follow Levins and Lewontin (1985) in framing this work as understood and articulated by a dialectical *primatologist*, rather than calling for a specific dialectical *primatology*. The former seeks to articulate general principles while simultaneously conceding variance, to be found and embraced, within individual practitioners. The latter is suggestive of a singular, coherent project or prescription, something that I am not advocating. The principles I will explore, with respect to primate research, conservation and ethics, with a focus on the apes, are encapsulated by the following themes: (1) *emergence*, or the dynamics among various levels of organisation (e.g., the individual, the group, the population) which are both semi-autonomous and reciprocally interacting; (2) *constrained totipotentiality*,[5] or the channelling of variation (including phenotypic flexibility) along pathways, and shaped by forces of the immediate, selective environment; and (3) *constructed compatibilities*, or the idea that in the formation of distinctions or dichotomies (e.g., research/conservation; captive/wild) there exists an inherent, non-trivial synthesis to be achieved.

The dialectical approach, where polar opposites are taken as inadequate, finds synthesis in seeing both the truth and the error in statements of thesis and antithesis.[6] As skilfully reviewed in John Bellamy Foster's (2000) *Marx's Ecology*, the dialectical tradition extends back to G.W.F. Hegel's idealist philosophy. At its core, a dialectical treatment attends to contradictions as a transcendent process of overcoming alienation and the severance of humans from the external world. Foster's impeccable scholarship demonstrates that Karl Marx, drawing on the materialist philosophy of Epicurus, emphasises the existence and ontological primacy of the external, physical world. In doing so, Marx's dialectics tacks away from the idealist philosophy of Hegel and into the realm of the material.

Moreover, Foster illustrates the influence of these philosophical positions, as well as the societal undercurrents of the mid-19th century, on Charles Darwin–Marx's materialist contemporary. Marx's worldview is grounded, literally, in the insights of Justus von Liebig and metabolic relationships (indeed, in capitalist agricultural "rifts") between humans and fundamental ecological processes. Darwin too had a long-standing interest in soils and soil altering agents, publishing *The Formation of Vegetable Mould, through the Actions of Worms, with Observations on Their Habits* in 1881, the year prior to his death (Kutschera & Elliott 2010). Darwin's materialism stems in part from a rejection of the static and mechanistic, natural theologic views of the time.[7] A fundamental parallelism between Marx and Darwin resides in anti-teleological views of society and nature, respectively, and a commitment to a materialist conception of history.[8]

Darwin (1859), who along with Alfred Russel Wallace (Darwin & Wallace 1858), articulated the foundational tenets of natural selection and the primacy of external, environmental forces in explaining evolution, managed to retain a role for both historical contingencies and reciprocal processes. Over time, however, reductionist trends in the wake of the Modern Synthesis (a coalescence of Mendelian genetics and adaptationist principles) resulted in the establishment of an evolutionary paradigm predicated upon the primacy of a reductionist version of Darwin and Wallace's overarching theory. In its most extreme form, sub-organismal determinants (i.e., allele variants) are modelled as powerful causal agents of higher-level arrangements (e.g., individuals within social groups). Important to my argument, however, are the strong challenges to this paradigm by Stephen Jay Gould, Richard Levins and Richard Lewontin. At the forefront of their influential contributions to evolutionary theory, is a re-invigoration of an explicit dialectical and materialist approach. In various works (e.g., Gould's (2002) *The Structure of Evolutionary Theory;* Gould & Lewontin's (1979) *The Spandrels of San Marco;* and Levins & Lewontin's (1985) *The Dialectical Biologist*), a collective attention on holism, historical contingency, constraint and reciprocal interactionism between organism and environment, effectively safeguards their approach from overly mechanistic or reductionist explanations, while retaining fidelity to Darwin's evolutionary framework.

A testament to the impact of Gould, Levins and Lewontin, and their collective critique of reductionism and hyper-adaptation, is the broader reach of their dialectical approach beyond the biological sciences, to achieve resonance and utility among social scientists. The critique is particularly useful when it is deployed against those in the social sciences that are susceptible to a kind of biological envy; especially when the version of biology they aspire to is already prone towards an over-reliance on over-deterministic physical forces (e.g., perceiving genes to function as "codes" or "blueprints")[9] (Lewontin & Levins 2007). Herein lies the danger of reductionism, and therefore the need for an alternative approach. Finding salience in the dialectical tradition, prominent environmental sociologists Richard York and Brett Clark, in reviewing the contributions of Gould, Levins and Lewontin, amplify calls for the rejection of deterministic approaches in favour of those that emphasise the roles of historical contingency and dynamic, emergent processes (Clark & York 2005; York & Clark 2005). Clark and York (2005: 21) write:

> A genuinely dialectical materialist position is fruitful for understanding the dynamic relationships that exist in nature and the evolutionary process of life. While social history cannot be reduced to natural history, it's a part of it. A dialectical stance is essential in order to understand the material world in terms of its own becoming: recognising that history is open, contingent and contradictory.

Of particular relevance to the present objective of understanding the past, present and future of human–ape interactions are the material impacts of human–animal relations writ-large. York and Longo (2017: 43) outline this fundamental premise in arguing that:

> Animals have influenced the course of history in many ways, seen and unseen, being central to the development of agriculture, the spread of zoonotic diseases, and the functioning of ecosystems in which humans are embedded. Likewise, humans have influenced the history of animals, affecting their evolution, the distribution of their populations, and the quality of their lives.

Human and animal inter-relationships have played a significant role in shaping the past and present of our shared existence. Given the close phylogenetic relations between ourselves and our fellow hominoids, a dialectical approach to the study of the shared hominoid niche requires consideration of internal similarities (as hominoids), as well as to external conditions of existence.[10]

I aim to add to the critique of the reductionist view of evolutionary biology as briefly summarised above. Aiding in this endeavour is the recognition that the landscape of evolutionary theory has shifted significantly since the aforementioned, influential works. Now 35 years on from *The Dialectical Biologist*, we have an array of major theoretical developments and refined research methods to assist us in our quest to understand ourselves and our closest biological relatives, as well as operationalising our role in shaping ecological systems. Indeed, we even have the recognition of a new geological epoch (the anthropocene) to facilitate new, and challenge old, assumptions about our place and role in the natural world (Crutzen & Stoermer 2000; Steffen et al. 2011; Ellis 2015). Fundamental to these new approaches and conceptual developments is the primacy of integrative niche dynamics (e.g., Odling-Smee et al. 1996; Laland et al. 2001; Fuentes 2015). In recognising the inseparability of the natural and the social, I see the potential for a dialectical approach to highlight the heterogeneity and complexity of ecological relationships. While reductionist and non-dialectical approaches in primatology provide utility in terms of hypothesis generation and model formation, they may veil fundamental biological and social interpenetrations. Relevant here is another component of Hegel's philosophy (and like dialectics, incorporated into Marxist thought) – and that is the concept of *totality*. Totality refers to "a methodological insistence that adequate understanding of complex phenomena can follow only from an appreciation of their relational integrity" (Jay 1984: 23–24). In combination, these approaches consider organisms as both subjects/objects and causes/effects within ecological niches. In primatology, such an approach can elucidate seemingly dichotomous categories, including: individual/group; intra-group/inter-group; and competition/cooperation.

To demonstrate the application of these themes to a primatological example, consider (briefly here, but in more detail in Chapters 2 and 3) the social systems of the small apes (family Hylobatidae). Despite their taxonomic diversity and broad geographic distribution, hylobatids were once considered to be invariably pair-bonded (in social structure) and monogamous (in mating pattern). In fact, a great deal of theorising and evolutionary importance was focused intensely on these particular facets of their social system. That is, to understand the fitness implications and underlying determinants of a particular social arrangement (i.e., the monogamous pair bond) was to understand the entirety of hylobatid sociality. This is an inherently reductionist approach, and its limitations were laid bare in 2009 with the release of *The Gibbons: New Perspectives on Small Ape Socioecology and Population Biology* (edited by Susan Lappan and Danielle J. Whittaker). By then, flexibility in gibbon socioecology was a clearly an emerging theme, and several contributors to this influential volume argued for substantial social and ecological variation within the family (e.g., Malone & Fuentes 2009; Reichard 2009).[11] A dialectical perspective requires us to consider: (1) the *emergence* of sociality within any given hylobatid *community*, where patterns of sub-adult migration, partner turnover events and extra-pair copulations create networks of social and genetic exchange at a level of organisation above that of the immediate social group; (2) the *particular histories* of exchange and demography that create an immediate selective environment (including existence and stability of greater-than-two-adult groups) that works with evolved behavioural patterns to *channel variation*; and (3) the re-conceptualisation of (previously) dichotomous distinctions (e.g., within-group as opposed to between-group behaviour) as important dialectical states for which a complete understanding of hylobatid sociality is now dependent. In theorising gibbon socioecology, a dialectical approach represents an integrative alternative to that of a non-dialectical, reductionist approach.

I will argue throughout the book that a dialectal analysis adds value to existing theoretical models and interpretive frameworks. However, it is within the realm of ape conservation (including the implementation of tactics and the associated ethics) where I deem a dialectical approach to be of the utmost necessity. We have a desperate need to refocus our efforts and increase our efficacy. Despite the dedicated efforts of numerous scientists and conservationists, we will likely witness the extinction of several ape species in the coming decades (Estrada et al. 2017; Estrada et al. 2020). How might interested scientists and conservation activists unite to stem the tide of humanity's impact? Even species that have received consistent attention are facing nearly insurmountable odds. To illustrate, by 2030, only 1% of Asian ape habitat will remain relatively undisturbed by human infrastructural and agricultural (e.g., oil palm) development (Nelleman & Newton 2002; Hockings et al. 2015). Across cultural, historical and ecological contexts, primates and humans share a coexistence (Tutin & Oslisly 1995; Sengupta & Radhakrishna 2018). At the same time, local people are often marginalised and exploited by conservation initiatives (Mulder & Coppolillo 2005; Brockington et al. 2006).

Ape conservation in the 21st century requires a synthesis of nature and culture that recognises their inseparability in ecological relationships that are both biophysically and socially formed.[12] Such a synthesis can emerge from the scholarly interrogation of dualisms that are deeply embedded within the intellectual traditions of the sciences and humanities (e.g., human/animal; nature/culture) (Haraway 2003; Fuentes 2010). The pronounced boundaries and divisions produced within and between scholarly fields have tended to dissociate humans and nature, sometimes to the point of exempting humans from basic physical properties and laws of nature (Catton & Dunlap 1978; York & Longo 2017). A key aim of this book is to draw upon my experiences in primate research and conservation in order to re-conceptualise the human–ape interface, and to develop culturally compelling conservation strategies required for the facilitation of hominoid coexistence. My overarching goal is not only to contribute to primatological theory, but to present a vision for our relationship with apes in the 21st century, and to ensure those relations extend into the 22nd (a premise that is not at all guaranteed).

My path as a dialectical primatologist

Here, I provide a brief overview of the various research projects that I have been involved with over the past 20+ years. In doing so, I establish the foundations for my thoughts on a variety of human–ape interfaces. Throughout the book, I draw on these findings and reflections, in combination with salient examples from the broader literature, to support the assertions herein. My very first research project, as an undergraduate in Marc Bekoff's animal behaviour course at the University of Colorado, was admittedly not very well designed nor well executed. However, the topic reveals an early interest in human–animal relations. I hypothesised that the duration of mutual eye contact, between humans and an array of captive animals at the Denver Zoo, should correlate with phylogenetic proximity between focal-species pairs. Despite confounds all over the show, I attribute this class project (along with Bekoff's engaging lectures on the behavioural interactions among dogs, and between dogs and their human companions) to generating some early insight into animal (and especially mammal) "ways of being" via evolved, and often shared, physiological and behavioural pathways. I further recall early encounters with debates in animal ethics and research, including the appropriate deployment of anthropomorphism (and the call to avoid "anthropodenial").[13] I continue to find such topics both fascinating and challenging.

After graduating from the University of Colorado with a major in anthropology, I spent approximately six months, first as an intern and then as a seasonal keeper, at the Brookfield Zoological Gardens in Chicago. Working alongside Rich Bergl as my mentor, I crafted an arboreal nesting platform out of donated lengths of fire hose, and observed the orangutans increase their locomotion in the upper portions of their night enclosures. From there, both Rich and I decided to pursue graduate-level studies. Rich travelled east to New York while I went west to the State of Washington. Upon arrival, my interest in structural-environmental enrichment for captive apes continued. Together with fellow graduate students John Mulcahy, Diana Goodrich and Quentin Davis, we constructed an array of structures that added vertical complexity to the enclosures at Central Washington University's Chimpanzee and Human Communication Institute (CHCI). Short write-ups of these projects in *The Shape of Enrichment*, a quarterly newsletter and source of ideas for behavioural enrichment, represent some of my first publications (Malone 1998; Malone 2000).

The chimpanzees at CHCI were remarkable beings. Washoe, Moja, Tatu, Dar and Loulis displayed all of the sentience, intelligence and emotional awareness found among all members of their species. They also possessed heightened capacities to communicate their interests and emotions to their human caregivers, and to each other, due to their unique history as participants in ape-language research.[14] At CHCI, I was privileged to spend a great deal of time around these extraordinary beings. Under the supervision of Drs Agustín Fuentes and Roger Fouts, and fellow graduate student Crickette Sanz, myself and others collected hundreds of hours of chimpanzee post-conflict data. After minor and sometimes major skirmishes among the chimpanzees, we recorded behaviour in the post-conflict and matched control periods. Importantly, we documented patterns and rates of reconciliation, redirection and consolation. Most interestingly, these findings revealed the ways in which human caregivers were co-opted into, and used effectively by the chimpanzees during, the post-conflict social milieu (Malone et al. 2000; Fuentes et al. 2002). By capturing this human element and making visible the impact of humans on the emerging social lives of the study participants, we joined the burgeoning movement within primatology that was problematising the barriers between an inescapable human presence and an objective lens on "natural" primate behaviour.[15]

At this same time, Agustín Fuentes was preparing me for my first trip to Indonesia to collect data for my masters-level qualification. The plan was for me to join with local primate researchers and survey public bird markets (or *pasar burung*) on the islands of Java and Bali to document the sale of primates therein. Despite their name, these markets are notorious for their long-standing presence and the sale (in many cases illegal) of a diverse array of Asian fauna.[16] With expert assistance from a flourishing community of wildlife researchers and activists from non-governmental organisations, such as KONUS in Bandung, West Java and KSBK (later ProFauna) in Malang, East Java, we managed to obtain an indication of the scope and scale of the illegal primate trade (Malone

Figure 1.1 Infant agile gibbon (*Hylobates agilis*) as presented for sale in the Pramuka Market, Jakarta, July 2000.

Photo credit: Meredith Malone.

et al. 2002; Malone et al. 2004) (Figure 1.1). One of the many lessons learned during that research was the absolute necessity of working openly and generously with local collaborators. We concluded:

> The collaboration and involvement of local peoples, including governmental and nongovernmental organisations, is essential for the effective monitoring of certain activities that have inherent connections to cultural and economic systems. Further investigations into the sellers' network and, ultimately, connections to primate source origins are needed to understand the economic and social implications of this trade, as well as the ecological implications for source populations.
>
> (Malone et al. 2004: 48)

In further market surveys, participation in confiscation activities, and in the conduct of behavioural studies of wild gibbons, the spectre of the primate trade is ever-present. In light of this, establishing ongoing collaborations and demonstrating a respect for local laws and customs are of paramount importance.

Over the next decade and a half, I would return frequently to Jakarta, occasionally revisiting the Pramuka and Barito bird markets, but most often to attend to the time-intensive process of research permitting, police

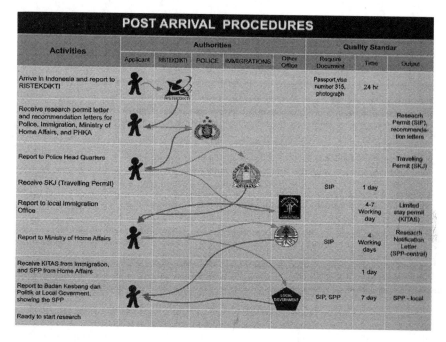

Figure 1.2 Flow chart of required post-arrival procedures, subsequent to receiving approval of the research permit.

Orignal image courtesy of the Republic of Indonesia's Ministry of Research, Technology and Higher Education (RISTEKDIKTI) Foreign Research Permit Guide.

registration and immigration paperwork required to legally conduct research in Indonesia.[17] It is not uncommon, especially during the North American summer, for primatologists to encounter one another along the convoluted path of "post-arrival procedures" (Figure 1.2). Exchanging tips and perhaps sharing resources, such as the elusive but necessary "revenue stamps" (*materai*), I often met colleagues for whom the ultimate destination, beyond the bureaucratic process, was either Sumatra, Kalimantan or further afield. It was there, but certainly not Java, where primate researchers went to observe Indonesia's wild primates within relatively healthy forests. In Java, the forests were too badly disturbed or destroyed, the human density too high, for studies in primate behavioural ecology to yield un-compromised, evolutionary insights. I, however, became intrigued with Java, and in particular West Java. For me, the island's central role in the nation's politics and economy, in combination with a rich cultural history and pockets of resilient primate communities, has proven to be the ultimate research setting.

As an anthropologist and primatologist, I felt compelled to understand the complex ecology and nuanced perceptions of primate life in Java. For my PhD, I began a long-term investigation of silvery gibbon population dynamics.

Over the years I would collect information on intra- and inter-group social behaviour, basic indicators of habitat quality and health, connections between wild populations and the primate trade, and ethnographic insights from an array of forest users (Malone 2007; Malone & Fuentes 2009; Malone et al. 2014; Malone & Wedana 2019). The forests of Southeast Asia, generally, and in West Java particularly, are powerful domains, in both a material and representational sense, where personal and collective histories, spiritualities, economies and political influences intersect.[18] Every research trip further uncovers additional nuances, humbling realisations, sharp contradictions and hints of new mysteries. In short, Java is a both formative and sustaining place for a dialectical primatologist.

After completing my doctorate, I had the privilege of joining a research team from the University of Oregon as a postdoctoral fellow on a project that took us into the heart of the Congo Basin rainforest. After many days of travel and expertly handled negotiations (or "argy bargy", as Frances White referred to it), we arrived in the truly magnificent Lomako Forest. For me, the experience was highly influential. Previous reading about the distinct social systems of bonobos (i.e., female bonding without female relatedness and limited bonding among related males) is one thing, but to observe this first-hand sets the gears of the socioecological theorist's mind in motion. How are we to assess the evolutionary trajectory of this system relative to the other African apes? What are the ecological underpinnings of the social system? How different are the organism–environment interactions between bonobos and chimpanzees? Furthermore, although nest counts and direct observation provided evidence of enhanced population productivity (White et al. 2008), how much of a buffer is afforded to a population that is under siege from unpredictable, external (and very human) forces? And finally, as previously mentioned, the ethical considerations and conservation benefits of the research were constantly at the forefront of our thinking.

My scholarly trajectory has been influenced beyond measure by my teachers and mentors. My training in primatology began with exposure to the basics of socioecological models from Professor Michelle Sauther (University of Colorado), as well as exposure to the wonderful diversity of extant and extinct prosimians, monkeys and apes. Subsequently, my early graduate school years were heavily influenced by Professor Agustín Fuentes (from the early days at Central Washington University, through his 17-year tenure at Notre Dame, and now presently at Princeton). Agustín's critical perspectives on overarching models and their requisite assumptions instilled in me the need to carefully evaluate (and sometimes challenge) the received primatological wisdom of the day. I learned that as an anthropologist, it is important to constantly think reflexively with respect to research design, implementation and the interpretation of salient data. Furthermore, the implications of research activity, both intended and otherwise, require additional attention. Later, while studying for my PhD at the University of Oregon (and as a postdoctoral fellow in the Democratic

Republic of Congo), I was the benefactor of insightful training from Professors Frances White and Joanna Lambert (now at the University of Colorado) in both primate ecology and conservation (theory and practice).

All of my aforementioned mentors provided formative contributions to my academic development, and these influences are clearly detectable in my past and present research engagements. With hindsight, the training I received from these academic mentors prepared me to effectively absorb and process the experiences and lessons that have accumulated thereafter, including: exposure to academic debates within university environments: the practical and ethical challenges of working as a foreign researcher in primate range countries; and encounters with displaced apes in a variety of settings. Lessons learned, both in the classroom and in the field, prepared me to embrace the ubiquitous presence and impact of humans within ecological systems, not as a hindrance, but as an opportunity to understand primate behaviour and ecology. This training also compels me to connect research results with data-driven conservation efforts as an obligation, not an option. In my mentors, the intellectual history of *their* supervisors is also represented, namely the late Robert Sussman, Phyllis Dolhinow, Paul Garber and John Fleagle. All of these scholars, through their dedication to the discipline have made notable contributions to research, conservation and capacity building within primate range countries. The impact and influence of these academic lineages on myself, and many other primatologists, cannot be overstated.[19]

Finally, in addition to my own research, I have been fortunate to supervise several very talented students since taking up my lectureship at the University of Auckland in 2010. My students have completed an array of relevant research, including: the socio-political and behavioural aspects of gibbon rehabilitation efforts (Alison Wade and Jen Hale for their MAs) (Wade 2012; Hale 2019); orangutan and human–caregiver interactions at the Auckland Zoo (Ally Palmer for her MA) (Palmer 2012; Palmer et al. 2015); the mediating role of technology in the experience of tourists to Bali's temple macaque sites (Kathryn Ovenden for her MA) (Ovenden 2015); human and orangutan coexistence in North Sumatra (Lotte van den Hout for her MA) (van den Hout 2020); and the nesting ecology of the Cross River gorilla and the Nigeria–Cameroon chimpanzee (Alison Wade again, this time for her PhD) (Wade et al. 2019; Wade 2020). In all of these projects, aspects of the human–nonhuman primate interface are further explored in a variety of contexts including the wild, the captive and the in-between. I will be referring to many of these studies in-detail in subsequent chapters. I am grateful to all my students for their efforts, and for the opportunity to derive additional and formative insights from my perspective as their supervisor. In sum, the path of the anthropological primatologist has been rewarding. While my experience in the discipline is particular, it is not unique. For colleagues, students and future researchers and/or conservationists, I hope this book will make it easier to unlock the full potential of primatology.

A brief overview of the book

In the present chapter, I set about a general discussion of dialectical principles and their utility in the primatological context. As the sub-title of the book indicates, I am particularly interested in the past, present and future of life in the hominoid niche, and how a dialectical approach to research and conservation may help to facilitate an informed and compassionate coexistence between humans and the other hominoid taxa (the apes). However, before I am able to dive deeper into these themes, some up-to-date, background information is required. In Chapter 2, "From the Miocene to the Margins", I provide a detailed summary of the superfamily Hominoidea. I begin this chapter by detailing the basic taxonomic relationships among the hominoids, as well as evolutionary timelines and trajectories of the major lineages, explaining how this once diverse group of primates journeyed from the mainstream of the Miocene to the margins of the present. Next, I will offer an overview of salient behavioural and ecological data from the hominoid taxa. From the relatively species–rich genera that comprise the Hylobatidae (small apes) to the handful of remaining species of extant Hominidae (great apes), including the subfamilies Ponginae and Homininae, I will compare and contrast the distinguishable features of each taxa. Finally, I will review (acknowledging that this is a rapidly developing field) the current conservation status of more than 25+ species (and even more taxa, subspecies inclusive) of living hominoids. All of these species with one exception (us) are faced with the threat of extinction. As our relationships with the other taxa that comprise the superfamily Hominoidea represent important markers in the navigation of humankind's relationship with the natural world, ensuring future connectivity and coexistence is a project of global, environmental importance. To start we will need to know where we presently stand.

In Chapter 3, "Emergence: Theorising Ape Sociality", I trace the history of primatological inquiry, from early expeditions and the collection of natural history data, through to a more deductive, systematic and hypothesis–driven science. Along the way theoretical models have been developed, tested and re-worked. I will review these theoretical developments with an emphasis on their strengths and shortcomings. Importantly, I will demonstrate the limitations of reductive models in the face of an emergent understanding of complex, primate social dynamics. It is here, critically, that three important tenets of my argument are established. First, the primacy of natural selection as an explanatory factor in evolutionary histories (a cornerstone of standard evolutionary theory, or SET) should be critically examined in light of recent developments. This re-examination is collectively referred to by proponents as the extended evolutionary synthesis (EES) (e.g., Laland et al. 2015; Müller 2017; Zeder 2017). One implication is that the social traits of organisms should not be considered *a priori* as fitness enhancing adaptations, but rather require empirical support to be classed as such.[20] Second, that given the parallel complexities of primate life histories and social lives (i.e., large-brained, slowly maturing mammals within a constantly shifting social milieu) we must attempt to model individual *behavioural flexibility* as a baseline capability.[21] This embrace of emergence and higher-level complexity forces us to view reductionist approaches with scepticism. And, finally, as primatologists come to terms with the ubiquitous presence of human activities, impacting to various degrees the lives and habitats of apes, our toolkits (theoretical and methodological) require some rethinking. Fortunately, this work is well underway (e.g., Fuentes 2012; Dore et al. 2017; Behie et al. 2019; Riley 2020).

If human-impacted localities typify primatology's new "sites of relevance", then arguably no other location is more relevant than Java. Java's distinctiveness emerges from its unique combination of biotic richness, physical geography and complex human history. In Chapter 4, "Waves of Change: Insights from Java, Indonesia", I delve into my experiences in this Indonesian island to demonstrate the complexities and realities of the human/nonhuman primate interface. Beginning with an initial assessment of the nonhuman primate trade in animal markets across Java (and Bali), and culminating with the long-term investigation of gibbon population dynamics, my research in Indonesia has involved (and indeed depended upon) a myriad of dedicated Indonesian researchers and conservationists. During my various projects, simultaneous attention to both socioecological data and ethnographic insights continue to reveal the complex ecologies and nuanced realities of primate life in Java. Understanding these interwoven layers, and working productively towards sustainable outcomes for both human and nonhuman primates, is the promise of an informed, anthropological primatology. While previously compartmentalised, the objectives of obtaining socioecological data and contributing to conservation measures have become integrated and mutually reinforcing over time, and are now bound by theoretical ties that attend to the entanglement of human and other primates' lives in the anthropocene.

In Chapter 5, "Betwixt and Between: Apes in (and on) the Verge", I present a variety of results and interpretations from my own studies, as well as research by my students and collaborators across a wide-array of research settings. From isolated wild populations, to highly managed captive groups, and the "gray sites" in between, I bring to light conflicts of ethics and efficacy in primate research and conservation.[22] From debates about the welfare of individuals versus the viability of populations and species, to the rights of researchers and local actors, this chapter will extend recent trends in primatology (e.g., the ethnoprimatological approach) to the increasingly "managed" human–ape interface. Are zoological gardens efficacious conservation partners? Should we direct our limited conservation resources towards habitat protection or towards more hands-on, animal-focused measures (e.g., confiscation of illegally-kept pets; rehabilitation and reintroduction)? Here again, the dialectical approach will be deployed in search of syntheses on topics for which polarising and polemical positions tend to dominate the debates. And, finally, in Chapter 6, "The Future of Life in the Hominoid Niche", I outline a holistic, synthetic vision for facilitating hominoid coexistence in the 21st century and beyond.

Primate research and conservation will certainly play a role, but the past decades have witnessed realignments and reconfigurations of these endeavours that previously were more clearly delineated (Behie et al. 2019). As any given primate community experiences anthropogenic alterations across a spectrum of types, frequencies and intensities (McKinney 2015; McLennan et al. 2017), I seek to identify general principles for coexistence rather than specific prescriptions. For example, incorporating the perceptions and motivations of stakeholders, be they conservation managers or local forest users, into ecological and behavioural analyses is a requisite step towards the creation of culturally compelling (and therefore viable) conservation initiatives.

To close this introductory chapter, allow me a statement of disclosure. When I proposed this book and its overarching theme (in early-2019), I wrote candidly of the fact that the direction of the work and its ability to chart a course towards the future of "hominoid coexistence" were neither established nor knowable at the outset. Rather, the synthetic set of principles that I endeavoured to propose would develop organically and as a result of the analytic work undertaken during the writing process. And the process has been as follows. First, I have attempted to understand how the observed patterns of variability within- and between-hominoid taxa are simultaneously shaped by, and act as shaping factors of, evolutionary processes. I have searched for examples of how apes, as long-lived and large-brained primates, are able to adjust their behaviour to minor shifts in ecological and demographic conditions via social innovation and learning (e.g., Campbell-Smith et al. 2011; Hockings & McLennan 2012). However, all hominoids (with the exception of humans) are also slow-reproducing and therefore particularly vulnerable to rapid environmental alterations, such as those brought about by humans. As such, what are the limits of ape resilience in the face of human activities? Second, I argue that as unsustainable human activities threaten ~60% of all primate species with extinction (and 100% of the apes), effective conservation strategies for these threatened apes will depend partially upon our understanding of behavioural responses to human-modified habitats, and partly upon our priorities and perceptions of the apes themselves. I seek to find answers to crucial questions along the path that leads to either hominoid coexistence or, alternatively, extinction. In the end, I hope the reader understands how a dialectical analysis assists us in assessing the myriad factors within complex human–ape interfaces, and, importantly, how we can use this knowledge to inform conservation theory and practice.

Notes

1   Many scholars of Indonesian cultural dynamics, but most notably Clifford Geertz, have written about the importance of cultivating and projecting a refined and civilised ethic of the self. In *The Religion of Java*, Geertz describes two important and inter-related pairs of concepts, the first of which being *alus* (*halus*, I.) and *kasar*. Geertz (1960: 232) writes:

Alus means pure, refined, polished, polite, exquisite, ethereal, subtle, civilised, smooth. A man who speaks flawless high-Javanese is *alus*, as is the high-Javanese itself … One's own soul and character are *alus* insofar as one emotionally comprehends the ultimate structure of existence; and one's behaviour and actions are *alus* insofar as they are regulated by the delicate intricacies of the complex court-derived etiquette. *Kasar* is merely the opposite: impolite, rough, uncivilised; a badly played piece of music, a stupid joke, a cheap piece of cloth.

A complete understanding of the *alus/kasar* dichotomy requires comprehension of a second pair of concepts: *batin* and *lair* (*lahir*, I.). Geertz (1960: 232) continues:

Batin means "the inner realm of human experience," and *lair* "the outer realm of human behaviour." … These two sets of phenomena, the inner and the outer, are conceived as somewhat independent realms to be put into proper order separately, or, perhaps better stated, sequentially. The ordering of the outward life leaves one free to turn to the ordering of the inward. The cultivated man needs to give form both to the naturally jagged physical gestures which make up his external behaviour and to the fluctuating states of feeling which comprise his inner experience. A truly *alus* man is polite all the way through.

2   In the introductory chapter to *Centralising Fieldwork*, MacClancy and Fuentes foreground the nature of fieldwork as both process and ethical practice. They note that "once in the field, there is no easy escape from ethical involvement" (2011: 21). These words ring especially true for researchers working in war-torn regions. With the support and participation of the local people (especially Papa Ikwa Nyamaolo Bosco, Papa Terrible, Papa Sayo and Papa Charmant) bonobo research at Lomako (and the main study sites of N'dele and Iyema) has continued under extremely challenging conditions. Our understanding of bonobos at these sites stems from the dedicated efforts of Drs Noel and Alison Badrian, Nancy Thompson-Handler, Richard Malenky, Frances White, Jef Dupain, Gottfried Hohmann, Barbara Fruth, Michel Waller, Amy Cobden and others. See also Waller and White (2016).

3   Researchers investigating primate behaviour, cognition, ecology and evolution can emerge from a number of cognate disciplines, including anthropology, biology and psychology. My personal academic background is in anthropology (for BA and PhD degrees) and psychology (with a specialisation in primate behaviour, for the MSc). Anthropology, as commonly constituted in the North American tradition, encompasses a multi-field approach that includes sociocultural anthropology, archaeology, biological anthropology and linguistic anthropology. Elsewhere (e.g., in the United Kingdom or in Europe) the various sub-fields of anthropology constitute separate departments with distinct intellectual histories. For graduate students in the multi-field tradition, training typically mandates breadth across the sub-disciplines, and often explicitly prepares students for the practical and ethical rigours of fieldwork. This training, I contend, better positions anthropologically trained primatologists, more so than those emerging from other scholarly traditions, for the inherent challenges of primate research and conservation. Though seemingly not definitive in its original attribution (Alfred Kroeber or Eric Wolf), the frequently bandied about statement that anthropology is "both the most scientific of the humanities and the most humanistic of the sciences" speaks to the position of the discipline. For a more comprehensive analysis of the science/humanities juxtaposition, see Lett (1997) and Riley (2020).

4 Gould (1989: 283–284) eloquently clarified the meaning of "contingent":

> I am not speaking of randomness, but of the central principle of all history – contingency. A historical explanation does not rest on direct deductions from laws of nature, but on an unpredictable sequence of antecedent states, where any major change in any step of the sequence would have altered the final result. This final result is therefore dependent, or contingent, upon everything that came before – the un-erasable and determining signature of history.

5 Here, I am borrowing and adapting the term *totipotentiality* from Robert Sussman. Sussman (2012: 185) declares: "the totipotentiality of behaviours for each species differs". The idea is that each species is unique with respect to the total potential of their behavioural repertoire (inclusive of varying degrees of overlap in cross-species comparisons). For species, like primates, with a high degree of intraspecific behavioural variation, *constrained totipotentiality* refers to how this variation plays out over both evolutionary and ecological timescales.

6 Here, I include an informative quote on dialectical principles by Friedrich Engels (1878/1969: 111) in *Anti-Dühring*:

> Truth and error, like all thought-concepts which move in polar opposites, have absolute validity only in an extremely limited field, as we have just seen, and as even Herr Dühring would realise if he had any acquaintance with the first elements of dialectics, which deal precisely with the inadequacy of all polar opposites. As soon as we apply the antithesis between truth and error outside of that narrow field which has been referred to above it becomes relative and therefore unserviceable for exact scientific modes of expression, and if we attempt to apply it as absolutely valid outside that field we really find ourselves altogether beaten: both poles of the antithesis become transformed into their opposites, truth becomes error and error truth.

7 Darwin's commitment to materialist views, in spite of the inevitable controversies, in effect lowered the status of humans by linking them to "lower" species while simultaneously raising the status of animals relative to humans (Foster 2000). Darwin (1859) famously concludes by seeing "grandeur" in the inter-connectedness of all life vis-à-vis a shared set of evolutionary principles. In contrast, see William Paley's (1803) *Natural Theology* for a discussion of nature's intricate "contrivances" as evidence for (and admiration of) the "consummate skill of the contriver".

8 See Gerratana (1973) for further reading on the relationship between Charles Darwin and Karl Marx, as well as insight into the varying influences of Thomas Malthus and Friedrich Engels on the scholarship of these contemporaries.

9 For a coherent, anti-reductionist case against the prevailing concept of the "gene as the primary causal agent in development", see Evelyn Fox-Keller's (2000) *The Century of the Gene*. Fuentes (2019: 4), referencing Fox-Keller, captures the more nuanced, contemporary view:

> DNA is part of an amazing, intricate system of interrelated proteins, enzymes, and other molecules and chemical relationships that interact to enable core aspects of development of organisms and their patterns of life. DNA cannot do anything alone, and it does not contain either the secret of life or a blueprint. It does offer us a great deal of information about life and its relationships.

10 In identifying similar themes inherent to the discipline of biological anthropology, Fuentes (2020: 3) states:

> There is a materiality of bodies and their interactions with one another and a physical manifestation of histories and patterns and processes of these interactions, but there is always more than the physical material involved in human experiences. Perceptions, ideologies, linguistic articulations, semiotic landscapes all matter in a serious understanding of the human. Humans are not unique in the world. That is, we are not wholly separate from other organisms, particularly the closely related other primates and many complex social mammals, who are also the targets of biological anthropological inquiry.

11 Fuentes (1998, 2000) undertook a re-evaluation of primate monogamy generally, and as it pertains to gibbons, specifically. Although a two-adult, pair-bonded group (along with one to two immature individuals) is still the most commonly observed gibbon social unit, significant variation is now documented for each and every aspect of the social system, including group composition, mating patterns and relationships within and between groups (Lappan & Whittaker 2009). The more recent perspective, that of emphasising the variable nature of hylobatid social systems, is not without its detractors, including prominent statements by gibbon expert David J. Chivers. It seems as if the objection arises, in part, from a place of reasoned caution regarding "the exceptions now being encountered more frequently and described more forcefully seem to me to be related clearly to isolated, disturbed, or fragmented habitats, and the various problems of overcrowding or imbalanced sex ratios associated with that" (Chivers 2009: vii). However, personal biases and cultural assumptions are also evident. For example, Chivers (2009: vi) confesses emphatically: "I live in hope that it will be shown eventually that we are descended from the lovely Asian apes, rather than those unattractive and promiscuous African apes with swollen bottoms!"

12 Here, and elsewhere in the book, I will deploy the term *natureculture*, a concept created by, and developed in Haraway's (2003) *The Companion Species Manifesto*.

13 Frans de Waal (1997) has asked the question: are we in *anthropodenial*?

14 For a comprehensive overview of this research, the chimpanzees and the Chimpanzee and Human Communication Institute, see Fouts and Mills (1997).

15 Leslie Sponsel (1997) is credited for the original use of the term "ethnoprimatology" in describing the complexity of the human niche within Amazonia, including human and nonhuman primate interactions. With the publication in 2002 of *Primates Face to Face* by Agustín Fuentes and Linda Wolf (editors), the study of human and nonhuman primate interactions moved onto primatology's main stage. See Riley (2018, 2020) for more on the emergence and maturational history of ethnoprimatology, including the acknowledgement that early attention was drawn to the near-ubiquity of human influence within ecosystems by Strum and Western (1982). Also, see Palmer and Malone (2018) for additional examples of the study of human and nonhuman primate interconnections in managed settings.

16 Describing the "accelerating bustle and pace of Batavian life", Winchester (2004: 141) wrote:

> As early as 1832, according to an account written by the renowned Dutch scholar of Javanese, Professor P.P. Roorda van Eysinga, there were widespread signs of an accelerating prosperity: ... Rich Arabs and Chinese rode through the

streets, half-clad Javanese carried heavy loads, shabby European clerks walked beneath sunshades to their offices, old women sold cakes, an Indian sat calmly eating his rice on a banana leaf, and vegetable, milk and fruit sellers, butchers and hill-dwellers offering monkeys and birds all mingled together in the crowd.

17 Associate Professor Sharyn Graham Davies, my colleague and Director of the Herb Feith Indonesian Engagement Centre at Monash University in Victoria, Australia, compiled a set of stories and reports from researchers with experience pursuing the permitting process in Indonesia. In short, those aspiring to legally conduct research in Indonesia should plan on spending (minimally) multiple months and significant financial resources to complete the various processes. See: www.academia.edu/40411588/Indonesia_Research_Visa_and_Research_Permit

18 See Anna Tsing's (2005) *Friction: An Ethnography of Global Connection*. Reflecting on her research in the Meratus Mountains of South Kalimantan, Tsing writes:

> The forest landscape is *social* [original emphasis]. I originally entered the Meratus forests with the eyes of a naturalist. I marvelled at the diversity of species, and I admired the forest views from many a mountain ridge. It was only by walking and working with Meratus Dayaks that I learned to see the forest differently. The forest they showed me was a terrain of personal biography and community history
>
> (2005: xi)

See also selective works by Robert Wessing (e.g., 1988, 1993).

19 For the network of "ancestor/descendant relationships" among biological anthropologists, especially as it demonstrates the influence of particular academic advisors, visit the Academic Phylogeny of Biological Anthropology (www.bioanthtree.org).

20 Traits of organisms may become more or less frequent within populations as a result of the various mechanisms of evolutionary change over generational time, including: gene flow, genetic drift and natural selection. It is prudent to not assume that natural selection, the only mechanism capable of producing adaptations, is the sole, or primary mechanism acting upon variation at any given time. Marks (2008) demonstrates the difficulties in teasing apart the relationships between traits and evolutionary mechanism for three diseases: sickle-cell anaemia, porphyria variegata and Tay-Sachs. For more specific examples of how all "adaptations" are not created equal, see Gould and Vrba (1982) on the topic of "exaptation" – where organisms may derive a present utility from traits that differs from a past utility.

21 Strier (2017b) clarifies oft-conflated language surrounding the usage of the terms behavioural flexibility as compared to behavioural variation. Strier (7) states:

> behavioural flexibility always involves a response to external stimuli, and therefore always has a temporal component … behavioural variation can also have a temporal component when it is the product of environmental selection pressures acting over generations. From this perspective, another key distinction of behavioural flexibility is that it occurs over an individual's lifetime, or ecological time, instead of over the generations involved with adaptations over evolutionary time. In other words, although all forms of behavioural flexibility involve some degree of behavioural variance, not all behavioural variance can necessarily be attributed to flexibility in an individual response.

22 Irus Braverman (2014) uses the term "gray sites" to describe the increasingly complicated and intertwined set of relations between captive and wild populations. As nature reserves and pockets of free-living animals require more intense management, the distinctions between concepts such as "*in-situ*"/"*ex-situ*" and "captive"/ "wild" become increasingly obscured.

References

Behie A, Teichroeb J, Malone N. (Eds.). 2019. *Primate Research and Conservation in the Anthropocene* (Vol. 82). Cambridge: Cambridge University Press.

Braverman I. 2014. Conservation without nature: The trouble with in situ versus ex situ conservation. *Geoforum* 51: 47–57.

Brockington D, Igoe J, Schmidt-Soltau KAI. 2006. Conservation, human rights, and poverty reduction. *Conservation Biology* 20(1): 250–252.

Campbell-Smith G, Cambell-Smith M, Singleton I, Linkie M. 2011. Apes in space: Saving an imperilled orangutan population in Sumatra. *PLoS ONE* 6(2): e17210.

Catton Jr. WR, Dunlap RE. 1978. Environmental sociology: A new paradigm. *The American Sociologist* 13: 41–49.

Chivers DJ. 2009. Foreword. In Whittaker D, Lappan S. (Eds.), *The Gibbons: New Perspectives on Small Ape Socioecology and Population Biology* (pp. v–vii). Series Title: Developments in Primatology: Progress and Prospects. New York, NY: Springer Academic Press.

Clark B, York R. 2005. Dialectical nature. *Monthly Review* 57(1): 13–22.

Corbey R. 2005. *The Metaphysics of Apes: Negotiating the Animal-Human Boundary*. Cambridge: Cambridge University Press.

Corbey R, Theunissen B. 1995. *Ape, Man, Apeman: Changing Views since 1600*. Evaluative proceedings of the symposium Ape, man, apeman: Changing views since 1600, Leiden, the Netherlands, June 28–July 1, 1993. Leiden: Department of Prehistory, Leiden University.

Crutzen PJ, Stoermer EF. 2000. The Anthropocene. *Global Change Newsletter* 41: 17–18.

Darwin C. 1859/1968. *The Origin of Species, by Means of Natural Selection or the Preservation of Favoured Races in the Struggle for Life*. London: Penguin.

Darwin C. 1881. *The Formation of Vegetable Mould, through the Actions of Worms, with Observations on Their Habits*. London: John Murray.

Darwin C, Wallace AR. 1858. On the tendency of species to form varieties; and on the perpetuation of varieties and species by natural means of selection. *Zoological Journal of the Linnean Society* 3(9): 45–62.

de Waal FBM. 1997. Are we in anthropodenial? *Discover* 18(7): 50–53.

Dore KM, Riley EP, Fuentes A. (Eds.). 2017. *Ethnoprimatology: A Practical Guide to Research at the Human-Nonhuman Primate Interface* (Vol. 76). Cambridge: Cambridge University Press.

Ellis EC. 2015. Ecology in an anthropogenic biosphere. *Ecological Monographs* 85(3): 287–331.

Engels F. 1878/1969. *Anti-Dühring*. Moscow: Progress Publishers.

Estrada A, Garber PA, Chaudhary A. 2020. Current and future trends in socio-economic, demographic and governance factors affecting global primate conservation. *PeerJ* 8: e9816.

Estrada A, Garber PA, Rylands AB, Roos C, Fernandez-Duque E, Di Fiore A, … & Rovero F. 2017. Impending extinction crisis of the world's primates: Why primates matter. *Science Advances* 3(1): e1600946.

Foster JB. 2000. *Marx's Ecology: Materialism and Nature*. New York, NY: Monthly Review Press.

Fouts R, Mills ST. 1997. *Next of Kin: What Chimpanzees Have Taught Me about Who We Are*. New York, NY: William Morrow.

Fox-Keller E. 2000. *The Century of the Gene*. Cambridge, MA: Harvard University Press.

Fuentes A. 1998. Re-evaluating primate monogamy. *American Anthropologist* 100(4): 890–907.

Fuentes A. 2000. Hylobatid communities: Changing views on pair bonding and social organization in hominoids. *Yearbook of Physical Anthropology* 43: 33–60.

Fuentes A. 2010. Naturalcultural encounters in Bali: Monkeys, temples, tourists, and ethnoprimatology. *Cultural Anthropology* 25: 600–624.

Fuentes A. 2012. Ethnoprimatology and the anthropology of the human-primate interface. *Annual Review of Anthropology* 41: 101–117.

Fuentes A. 2015. Integrative anthropology and the human niche: Toward a contemporary approach to human evolution. *American Anthropologist* 117(2): 302–315.

Fuentes A. 2019. *Why We Believe: Evolution and the Human Way of Being*. New Haven, CT: Yale University Press.

Fuentes A. 2020. Biological anthropology's critical engagement with genomics, evolution, race/racism, and ourselves: Opportunities and challenges to making a difference in the academy and the world. *American Journal of Physical Anthropology* 175(2): 326–338.

Fuentes A, Malone N, Sanz C, Matheson MD, Vaughan L. 2002. Conflict and post-conflict behavior in a small group of chimpanzees (*Pan troglodytes*). *Primates* 43(3): 223–235.

Fuentes A, Wolfe LD. (Eds.). 2002. *Primates Face to Face: The Conservation Implications of Human-Nonhuman Primate Interconnections*. Cambridge: Cambridge University Press.

Geertz C. 1960. *The Religion of Java*. Chicago, IL: University of Chicago Press.

Gerratana V. 1973. Marx and Darwin. *New Left Review* 82: 60–82.

Gould SJ. 1977. *Ontogeny and Phylogeny*. Cambridge, MA: Harvard University Press.

Gould SJ. 1989. *Wonderful Life, the Burgess Shale and the Nature of History*. New York, NY: W.W. Norton & Company.

Gould SJ. 2002. *The Structure of Evolutionary Theory*. Cambridge, MA: Harvard University Press.

Gould SJ, Lewontin RC. 1979. The spandrels of San Marco and the Panglossian paradigm: A critique of the adaptationist programme. *Proceedings of the Royal Society of London. Series B. Biological Sciences* 205: 581–589.

Gould SJ, Vrba ES. 1982. Exaptation – a missing term in the science of form. *Paleobiology* 8(1): 4–15.

Hale J. 2019. *Within-Group Social Dynamics in Captive and Rescued Javan Gibbons (Hylobates moloch): Alternative Groups with Alternative Behaviours*. MA portfolio, University of Auckland, Auckland, New Zealand.

Haraway DJ. 1989. *Primate Visions: Gender, Race, and Nature in the World of Modern Science*. New York, NY: Routledge.

Haraway DJ. 2003. *The Companion Species Manifesto: Dogs, People, and Significant Otherness* (Vol. 1). Chicago, IL: Prickly Paradigm Press.

Hockings KJ, McLennan MR. 2012. From forest to farm: Systematic review of cultivar feeding by chimpanzees – Management implications for wildlife in anthropogenic landscapes. *PLoS ONE* 7(4): e33391.

Hockings KJ, McLennan MR, Carvalho S, Ancrenaz M, Bobe R, Byrne RW, ... & Wilson ML. 2015. Apes in the Anthropocene: Flexibility and survival. *Trends in Ecology & Evolution* 30(4): 215–222.

Jay M. 1984. *Marxism and Totality: The Adventures of a Concept from Lukács to Habermas.* Berkeley, CA: University of California Press.

Kutschera U, Elliott JM. 2010. Charles Darwin's observations on the behaviour of earthworms and the evolutionary history of a giant endemic species from Germany, *Lumbricus badensis* (Oligochaeta: Lumbricidae). *Applied and Environmental Soil Science* 2010: 823047.

Laland KN, Odling-Smee FJ, Feldman MW. 2001. Niche construction, ecological inheritance, and cycles of contingency in evolution. In Oyama S, Griffiths PE, Gray RD. (Eds.), *Cycles of Contingency: Developmental Systems and Evolution* (pp. 117–126). Cambridge, MA: MIT University Press.

Laland KN, Uller T, Feldman MW, Sterelny K, Müller GB, Moczek A, ... & Odling-Smee J. 2015. The extended evolutionary synthesis: Its structure, assumptions and predictions. *Proceedings of the Royal Society B: Biological Sciences* 282(1813): 20151019.

Lappan S, Whittaker DJ. (Eds.). 2009. *The Gibbons: New Perspectives on Small Ape Socioecology and Population Biology.* New York, NY: Springer.

Lett JW. 1997. *Science, Reason, and Anthropology: The Principles of Rational Inquiry.* Lanham, MD: Rowman & Littlefield.

Levins R, Lewontin R. 1985. *The Dialectical Biologist.* Cambridge, MA: Harvard University Press.

Lewontin R, Levins R. 2007. *Biology Under the Influence: Dialectical Essays on Ecology, Agriculture and Health.* New York, NY: Monthly Review Press.

MacClancy J, Fuentes A. (Eds.). 2011. *Centralizing Fieldwork: Critical Perspectives from Primatology, Biological, and Social Anthropology.* New York, NY: Berghahn Books.

Malone N. 1998. Providing orangutans with opportunities for arboreal behavior. *The Shape of Enrichment* 7(4): 1–2.

Malone N. 2000. Providing arboreal enrichment for captive chimpanzees. *The Shape of Enrichment* 9(2): 3–4.

Malone N. 2007. *The Socioecology of the Critically Endangered Javan Gibbon (Hylobates moloch): Assessing the Impact of Anthropogenic Disturbance on Primate Social Systems.* PhD Dissertation, University of Oregon, Eugene, OR.

Malone N, Fuentes A. 2009. The ecology and evolution of hylobatid communities: Proximate and ultimate considerations of inter- and intraspecific variation. In Whittaker D, Lappan S. (Eds.), *The Gibbons: New Perspectives on Small Ape Socioecology and Population Biology* (Chp. 12, pp. 241–264). Series Title: Developments in Primatology: Progress and Prospects. New York, NY: Springer Academic Press.

Malone N, Fuentes A, Purnama AR, Adi Putra IMW. 2004. Displaced hylobatids: Biological, cultural, and economic aspects of the primate trade in Java and Bali, Indonesia. *Tropical Biodiversity* 40(1): 41–49.

Malone N, Purnama AR, Wedana M, Fuentes A. 2002. Assessment of the sale of primates at Indonesian bird markets. *Asian Primates* 8(1–2): 7–11.

Malone N, Selby M, Longo SB. 2014. Political and ecological dimensions of silvery gibbon conservation efforts: An endangered ape in (and on) the verge. *International Journal of Sociology* 44(1): 34–53.

Malone N, Vaughan LL, Fuentes A. 2000. The role of human caregivers in the post-conflict interactions of captive chimpanzees (*Pan troglodytes*). *Laboratory Primate Newsletter* 39(1): 1–3.

Malone N, Wedana M. 2019. Struggling for socio-ecological resilience: A long-term study of silvery gibbons (*Hylobates moloch*) in the fragmented Sancang Forest Nature Reserve, West Java, Indonesia. In Behie A, Teichroeb J, Malone N. (Eds.), *Primate*

Research and Conservation in the Anthropocene (pp. 17–32). Cambridge: Cambridge University Press.

Marks JM. 2008. Would Darwin recognise himself here? In Elton S, O'Higgins P. (Eds.), *Medicine and Evolution: Current Applications, Future Prospects* (pp. 273–288). Boca Raton, FL: CRC Press.

Marks JM. 2011. *The Alternative Introduction to Biological Anthropology*. Oxford: Oxford University Press.

McKinney T. 2015. A classification system for describing anthropogenic influence on nonhuman primate populations. *American Journal of Primatology* 77(7): 715–726.

McLennan MR, Spagnoletti N, Hockings KJ. 2017. The implications of primate behavioral flexibility for sustainable human–primate coexistence in anthropogenic habitats. *International Journal of Primatology* 38(2): 105–121.

Mulder MB, Coppolillo P. 2005. *Conservation: Linking Ecology, Economics, and Culture.* Princeton, NJ: Princeton University Press.

Müller GB. 2017. Why an extended evolutionary synthesis is necessary. *Interface Focus* 7(5): 20170015.

Nelleman C, Newton A. 2002. Great apes – The road ahead. An analysis of great ape habitat, using GLOBIO methodology. *United Nations Environment Programme*, Nairobi.

Odling-Smee FJ, Laland KN, Feldman MW. 1996. Niche construction. *The American Naturalist* 147(4): 641–648.

Ovenden K. 2015. *Exploring the Tourist-Macaque Interface in Monkey Forests in Bali, Indonesia*. MA Portfolio, University of Auckland, New Zealand.

Palmer A. 2012. *Keeper/Orangutan Interactions at Auckland Zoo: Communication, Friendship and Ethics between Species*. MA Thesis, University of Auckland.

Palmer A, Malone N, Park J. 2015. Accessing orangutans' perspectives: Interdisciplinary methods at the human/animal interface. *Current Anthropology* 56(4): 571–578.

Paley W. 1803. *Natural Theology*. London: R. Faulder.

Reichard UH. 2009. The social organization and mating system of Khao Yai whitehanded gibbons: 1992–2006. In Whittaker D, Lappan S. (Eds.), *The Gibbons: New Perspectives on Small Ape Socioecology and Population Biology* (Chp. 12, pp. 347–384). Series Title: Developments in Primatology: Progress and Prospects. New York, NY: Springer Academic Press.

Riley EP. 2018. The maturation of ethnoprimatology: Theoretical and methodological pluralism. *International Journal of Primatology* 39(5): 705–729.

Riley EP. 2020. *The Promise of Contemporary Primatology*. New York, NY: Routledge.

Sengupta A, Radhakrishna S. 2018. The hand that feeds the monkey: Mutual influence of humans and rhesus macaques (*Macaca mulatta*) in the context of provisioning. *International Journal of Primatology* 39(5): 817–830.

Sponsel LE. 1997. The human niche in Amazonia: Explorations in ethnoprimatology. In Kinzey WG. (Ed.), *New World Primates: Ecology, Evolution, and Behavior* (pp. 143–165). New York, NY: American Anthropological Association.

Steffen W, Grinevald J, Crutzen P, McNeill J. 2011. The Anthropocene: Conceptual and historical perspectives. *Philosophical Transactions of the Royal Society of London A: Mathematical, Physical and Engineering Sciences* 369(1938): 842–867.

Strum SC, Western JD. 1982. Variations in fecundity with age and environment in olive baboons (*Papio anubis*). *American Journal of Primatology* 3(1–4): 61–76.

Sussman RW. 2012. Why we are not chimpanzees? In Calcagno JM, Fuentes A, What makes us human? Answers from evolutionary anthropology, p. 185, *Evolutionary Anthropology: Issues, News, and Reviews* 21(5): 182–194.

Tsing A. 2005. *Friction: An Ethnography of Global Connection*. Princeton, NJ: Princeton University Press.

Tutin CEG, Oslisly R. 1995. *Homo, Pan* and *Gorilla*: Co-existence over 60,000 years at Lope in Central Gabon. *Journal of Human Evolution* 28: 597–602.

Van den Hout LM. 2020. *An Ethnoprimatological Assessment of the Orang-utan/Human Landscape in North Sumatra Province, Indonesia*. MA Portfolio, University of Auckland, New Zealand.

Wade AH. 2012. *Exploring Ethnoprimatology*. MA Portfolio, University of Auckland, New Zealand.

Wade AH. 2020. *Shared Landscapes: The Human-Ape Interface within the Mone-Oku Forest, Cameroon*. PhD Thesis, University of Auckland, New Zealand.

Wade A, Malone N, Litttleton J, Floyd B. 2019. Uneasy neighbours: Local perceptions of the Cross River gorilla and Nigeria-Cameroon chimpanzee in Cameroon. In Behie A, Teichroeb J, Malone N. (Eds.), *Primate Research and Conservation in the Anthropocene* (pp. 52–73). Cambridge: Cambridge University Press.

Waller MT, White FJ. 2016. The effects of war on bonobos and other nonhuman primates in the Democratic Republic of the Congo. In Waller MT. (Ed.), *Ethnoprimatology: Primate Conservation in the 21st Century* (pp. 179–192). New York, NY: Springer.

Wessing R. 1988. Spirits of the earth and spirits of the water: Chthonic forces in the mountains of West Java. *Asian Folklore Studies* 47(1): 43–61.

Wessing R. 1993. A change in the forest: Myth and history in West Java. *Journal of Southeast Asian Studies* 24(1): 1–17.

White FJ, Waller MT, Cobden AK, Malone NM. 2008. Lomako bonobo population dynamics, habitat productivity, and the question of tool use. *American Journal of Physical Anthropology* 135(S46): 222.

York R, Clark B. 2005. Review essay: The science and humanism of Stephen Jay Gould. *Critical Sociology* 31(1–2): 281–295.

York R, Longo SB. 2017. Animals in the world: A materialist approach to sociological animal studies. *Journal of Sociology* 53(1): 32–46.

Zeder MA. 2017. Domestication as a model system for the extended evolutionary synthesis. *Interface Focus* 7(5): 20160133.

2 From the Miocene to the margins

Overview of the superfamily Hominoidea

> Hominoids, dominant for millions of years in a stable forest milieu, enjoyed almost runaway selection for cognition, learning, and extended life histories – all luxuries of a certain independence of the predator-prey cycle. In the face of a constrained milieu and unable to undo their life history pattern, hominoids evolved as expected: some lineages went extinct, some linger in the fragile remains of the original niche, and one abandoned its biological capacity completely in favour of its cognitive potential.
>
> (Walter Hartwig 2011: 31)

In addition to ourselves, the other hominoids or "apes" (members of the superfamily Hominoidea) – well known by their common names (the gibbons, siamang, orangutan, gorilla, chimpanzee and bonobo) – retain prominent positions in scientific and philosophical discussions of classification, order and hierarchy. For hundreds of years, apes have captured the attention of researchers, philosophers, zoologists and anatomists; for millennia they have been ascribed meaning by co-habitants in shared landscapes. For example, in Ancient and Imperial China, representations of gibbons within dynastic art reveal deep, symbolic linkages to gibbons that continue until present (Geissmann 2008). That is, the gibbon occupies a special niche in Chinese culture as a powerful and mysterious connective force between humankind and nature (Van Gulik 1967; Chatterjee 2016).[1] Coming to terms with our place among our biological next-of-kin is critical, not just for their future but for ours. Indeed, whereas the hominoid evolutionary trajectory extends back nearly 30 million years, "*Homo sapiens* is nowhere near a million years old and has limited prospects, entirely self-imposed, for extended geological longevity" (Gould 1994: 19).

An anthropological approach to evolution, in the context of continuity with the living apes, compels us to re-interpret and re-align the enigmatic boundaries between animal and human; and between nature and society. Raymond Corbey (2005: 91) captures this dialectic in describing:

> Human evolution scenarios, "up from the ape", inspired by an even more fundamental narrative on natural order, articulated human identity (who

DOI: 10.4324/9780367211349-2

we are) in terms of apish origins (where we come from). Metaphysically and morally authoritative stories were told and naturalised, helping to ward off threats and deal with what was emotionally disturbing and cognitively confusing.

Here, Corbey makes two noteworthy contributions. The first is to reinforce the notion that studies of human and ape evolution are inherently cultural endeavours. Second, and perhaps more importantly, the act of theorising ape evolution is not dissimilar to evolutionary processes themselves. That is to say, all are characterised by the inseparability of internal and external factors; dynamic, historical processes subject to constraint and contradiction. The dominant mode of analysis in the evolution of ape sociality, for example, centres on a framework whereby social systems are grounded in the distribution of risks and resources in the materiality of the niche (e.g., Wrangham 1986; Sterck et al. 1997). Yet, in these socioecological models the units of analysis we establish and measure, and subsequently treat as independent variables, are in reality more fluid, heterogeneous and inter-relational (see Chapter 3). Apes, like many other organisms, "both make and are made by the environment and are thus actors in their own evolutionary history" (Levins & Lewontin 1985: 274). A dialectical view embraces the interaction of what were once seen as mutually exclusive entities: organisms and environments; random and deterministic processes. When applied broadly to the evolution of hominoids, a dialectical approach brings us closer to an understanding of the subject in its full dimensionality.

So who exactly are the apes? How are their evolutionary trajectories distinct from the other primates, and how are they similar? What was the impetus for the evolution of the basic, ape "life-way", and how have humans extended this trajectory? What are the present threats to ape populations, and what are the limits of ape resilience in the face of humankind's expanding reach? In this chapter, I will attempt to answer these questions and synthesise our present knowledge of the apes. After providing an evolutionary summary of the superfamily Hominoidea, I will then present a comparative overview of extant ape diversity (a composite measure of species richness and abundance), social behaviour and ecology. As all of the hominoid species (with the exception of humans) are threatened with extinction, I will conclude by painting a general picture of challenges to ape conservation. The foundations established in this chapter provide critical infrastructure as I construct my case for the facilitation of human–ape coexistence in later stages of the book.

Miocene origins: Eco-morphological context

In tackling the question of ape-origins, a number of challenges endure, including the incompleteness of the fossil record and the inherent complexities of behavioural and ecological reconstructions. Fortunately, a number of prominent scholars have taken up the challenge and collectively advanced our knowledge. From John Napier's (1970) early chapter on paleoecology and catarrhine evolution, to Temerin and Cant's (1983) attempt to elucidate the cercopithecoid-hominoid divergence (i.e., Old World monkey-ape divergence), to more recent, comprehensive reviews (e.g., Hunt 2016; Begun 2017; Harrison 2017a; Andrews 2020), there are both promising leads and persisting challenges in attempts to "disentangle the diversity of Miocene apes in order to see the trajectory that led to the evolution of the living Hominoidea" (Begun 2017: 4). Although still unclear in a number of details, and always subject to new discoveries and analyses, one point that is no longer debated is that apes, once dominant and diverse in paleotropical habitats throughout the Miocene (23.0–5.0 million years ago [mya]), are now represented by but a shadow of their former selves. In contrast, the cercopithecoid monkeys, with humble beginnings and relatively limited diversity during the Miocene, underwent a more recent Plio–Pleistocene (5.0–1.8 mya) radiation resulting in a current position of strength (in both species richness and geographical distribution) relative to the living apes (Figure 2.1).[2]

Consistent with general trends in the evolution of the order Primates (Martin 1986; Sussman 1991; Sussman et al. 2013), the organisms we know today as the apes trace their evolutionary origins to the complex environs of tropical and sub-tropical forests. In the Oligocene/Miocene transition (approximately 25.0 mya), a warming climatic trend coupled with an increasing distribution of heterogeneous forest and woodland habitats provided a suitable context for emerging radiations of, initially medium-sized, arboreal catarrhine (paleotropical anthropoid) primates: the Hominoidea (apes) and the Cercopithecoidea (monkeys). These relatively aseasonal forests provided a consistent supply of readily available energy in the form of ripe fruit. For the emerging ape clade, this stability paved the way for an adaptive life-way that included protracted life histories and post-cranial specialisations (e.g., arboreal bridging and suspension; forelimb mobility and propulsion) that ensured efficient access to the

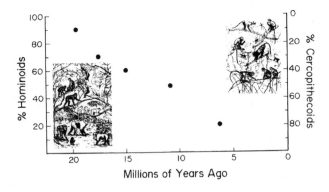

Figure 2.1 Analysis of the past 20 million years reveals a pattern of relative species diversity of hominoids and cercopithecoids: monkey diversity has increased whereas hominoid diversity has declined (Andrews 1986; Fleagle 1999; Harrison 2010; Hunt 2016).

high-quality diet on offer (Temerin & Cant 1983; Kelley 1997; Jablonski 2005). It is hypothesised that the ape evolutionary lineage "stays the course", so to speak, throughout the duration of the Miocene despite an eventual downward trend in the density and distribution of fruit patches. In most hominoid taxa (excepting the gibbons), the evolution of larger body size may be related to energy conservation, metabolic dynamics and an advantage in direct, competitive feeding bouts with smaller-bodied frugivores (Kleiber 1975; Hunt 2016). The evolution of more terrestrial locomotor repertoires (excepting the Asian apes) can also be seen as an energetic response to more patchily distributed feeding resources. The decline in the number of ape taxa in the late-Miocene can be attributed to the role of evolutionary constraints operating under conditions of diminishing, ecological opportunity.

During the same time period, paleotropical monkeys (superfamily Cercopithecoidea) were beginning an evolutionary journey of their own. Paleotropical monkeys feature an adaptive package that includes both specialisations (e.g., bilophodont molars, in combination with either cheek pouches or forestomach fermentation capabilities) and the retention of unspecialised traits (e.g., a postcranial anatomy suited for both arboreal and terrestrial, quadrupedal locomotion). Additionally, higher rates of intrinsic natural increase allow cercopithecoid species to rebound quickly from population-level declines (Jablonski 2005). Following the middle Miocene climatic optimum, forest habitats became drier and more seasonal. The diversity of the paleotropical monkeys, in terms of both geographical distribution and breadth of ecological niche occupancy, is testament to the overall strength of this primate radiation (Frost 2017). Ultimately, the varying evolutionary trajectories of paleotropical monkeys and apes, under ecologically similar conditions, established a terminal-Miocene baseline for the fates of these lineages in the Pliocene, Pleistocene, Holocene and beyond.

In order to get a sense for the diversity of ape species that have populated our past, here I present a brief (and non-comprehensive) overview of the fossil apes. The fossil record of the superfamily Hominoidea is characterised by a putative emergence of the taxon in East Africa during the earliest Miocene.[3] Although subject to some debate, the genus *Proconsul* is represented by as many as six species from early-to-middle Miocene deposits in Kenya and eastern Uganda, and were medium-large (10–90 kg), fruit eating, arboreal quadrupeds (Fleagle 1999; Harrison 2017b) (Figure 2.2). With a combination of derived dental features relative to early catarrhine primates[4] and a sufficiently generalised postcranial anatomy, *Proconsul spp.*, and an array of similar taxa from this time (e.g., *Ekembo*; *Morotopithecus*), possess a mixture of traits, at the right place and time, to be stem-Hominoids.[5] However, their precise phyletic position in the taxonomy of extinct and extant apes remain unclear, especially since many species possess a mosaic of primitive and derived features. Beginning at around 17 mya, hominoids were able to extend their range to Eurasia (as part of a more general pattern of land mammal dispersals) via intermittent land connections between Africa and Eurasia (Begun et al. 2012). Candidate lineages for such

Figure 2.2 Fossil ape community reconstruction from Rusinga Island, Kenya, depicting multiple taxa from the family Proconsulidae (Fleagle 1999).

a dispersal scenario can be found in: (a) the *Proconsul*-like *Afropithecus* (found in Kenya, during the latter parts of the early Miocene), which possessed a more robust chewing morphology and dental adaptions indicative of increased hard-object feeding, and putative continuity with *Heliopithecus* on the Arabian Peninsula (Andrews & Martin 1987; Fleagle 1999; Begun 2017); and (b) the presence of *Kenyapithecus* in both Kenya and Turkey, a core component of a trans-migratory radiation of thickly enamelled hominoids between roughly 17.0 and 13.0 mya (Begun et al. 2012).

The middle Miocene (ca. 16.0–10.0 mya) record preserves a diversity of forms and documents a vast geographic expansion. In Africa, and primarily Kenya, taxa such as *Nacholapithecus* hint at the development of modern-ape forelimb morphology, but lack the derived features common among all the living apes (Harrison 2017a). As African forests become drier and more sea-sonal, fossil ape diversity declines. However, it is in Europe and Asia where the diversity of middle Miocene hominoids is well-and-truly on display. Once well established to the north of the intercontinental land bridge, hominoids flourished in woodlands that stretched from southern Europe to south-western China. Importantly, we see the emergence of a hominoid postcranial anatomy that includes a suite of adaptations related to the stability of fore-limb joints through a range of suspensory and climbing motions. Indeed, the earliest members of the modern great ape subfamilies Ponginae and Homininae emerged almost simultaneously at ca. 12.5 mya in Asia and Europe, respect-ively. The relatively widespread distribution in western Europe of *Dryopithecus spp.*, and the intriguing yet more restricted *Pierolapithecus catalaunicus* in Spain, heralds the arrival and radiation of "the distinctive orthograde body plan of extant great apes" (Moyá-Solá 2017: 1). In these species, postcranial elements indicate a postural and locomotor anatomy that is derived relative any of the earlier hominoids (e.g., proconsulids), and indicative of the type of specialised vertical climbing and (perhaps even) suspensory capabilities present in all living apes. Yet, a mosaic of derived and primitive traits in these taxa, especially in the case of *Pierolapithecus* with its clear lack of any below-branch suspensory adaptations (but see Deane & Begun 2008), compels us to remain conservative in the drawing of conclusive, phyletic linkages (Pina et al. 2014).

At approximately the same time (12.5-7.5 mya), we see the emergence of a particularly interesting genus, *Sivapithecus*. *Sivapithecus* is known from multiple localities, and as such occupied varied habitats in (primarily, but not limited to) South Asia (Pakistan and India) (Andrews 1983). Due to striking craniofacial similarities between the extinct, Miocene-era *Sivapithecus* and the living orang-utan (*Pongo*), the former has long been considered a direct ancestor of the latter (Figure 2.3). However, *Sivapithecus* provides us with an insightful example of how the science of hominoid evolution is a fluid matter, subject to new discov-eries and interpretations. In the description and analysis of bones of the forelimb (humeri), Pilbeam et al. (1990) argue that *Sivapithecus* displayed characteristics expected of a general quadruped (as opposed to the suspensory postural and locomotor morphology shared among all the living hominoids). Following on from that discovery are two alternative interpretations, either: (a) the presumed close affinity between *Sivapithecus* and *Pongo* as sister taxa is unsupported; or

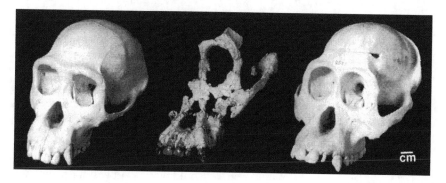

Figure 2.3 A comparison of the cranial remains of *Sivapithecus indicus* (GSP 15000) with the crania of both *Pan* (left) and of *Pongo* (right).
From Fleagle (1999) with photo courtesy of David Pilbeam.

(b) *Sivapithecus* and *Pongo* are sister taxa, and thus the many postcranial, suspensory attributes shared among the extant hominoids are the result of parallel evolution. Larson (1998: 87–88) summarises the ensuing challenge in that "either interpretation implies a significant amount of homoplasy, or similarities in two or more taxa brought about by independent acquisition of traits rather than through shared descent, during the evolutionary history of the hominoids". The situation may be further complicated by potential species-level diversity within the *Sivapithecus* genus – with both adaptations for either climbing and suspension or quadrupedal locomotion, in *S. indicus* and *S. parvada*, respectively (Moyá-Solá & Köhler 1996). One last and less controversial fact about *Sivapithecus* is that it is likely to be the direct ancestor of multiple species attributed to the genus *Gigantopithecus* – the largest primates that ever lived. *Gigantopithecus blacki*, at an estimated 300 kg, existed on a diet of hard and fibrous material, and survived until relatively recently (ca. 400,000 years ago) in Southeast Asia.

Further uncertainty exists with respect to the evolutionary linkages between middle-to-late Miocene apes and the living African ape genera. The once flourishing radiation of hominoids during the middle Miocene period in Europe began to decline after 10.0 mya due to increasing cooling and drying trends and the subsequent intensification of seasonal variability (Potts 2004; Begun 2017). Few candidate species demonstrate the right balance of primitive and derived traits to be considered putative members of the subfamily Homininae, or "hominines". One such form, especially with respect to facial morphology, is *Ouranopithecus* (sometimes referred to as *Graecopithecus*), a late Miocene ape from fossil assemblages in Greece (Dean & Delson 1992; Koufos & de Bonis 2005). Another fossil hominoid with distinctive affinities to the African apes is *Samburupithecus* (ca. 9.5 mya, Kenya) (Ishida & Pickford 1997). However, an alternative view based on the retention of primitive characteristics is that *Samburupithecus* is a late-surviving proconsulid (Begun 2007).

Despite the difficulties inherent in these reconstructions, what is almost certain is that the exchange of hominoids between Africa and Eurasia was unlikely exclusively uni-directional (north–south), but rather consisted of a minimum two dispersal events – one *out-of-*, and a second *into-*Africa (Disotell & Stewart 1998). From an early African stock, hominoids followed the northward spread of forested and woodland habitats in the middle Miocene. These colonising hominoids underwent diversification in Eurasia, and some dispersed eastward via favourable habitat corridors to easternmost Asia. Along that journey, many taxa evolved and subsequently went extinct. The living descendent taxa of this radiation are known to us today as the small apes (Hylobatidae) and the great apes, including humans (Hominidae). Genetic evidence points to a mid–Miocene date of between 18.5 and 15.0 mya for the Hylobatidae-Hominidae divergence (Tyler 1993; Raaum et al. 2005). With the earliest appearance of hominoids in Eurasia at ca. 16 mya, this early divergence would place a last common ancestor with the Hominidae on the African continent (or very soon after the aforementioned northward movement of hominoids into Eurasia). Unfortunately, the fossil record is nearly silent for the first 9–10 million years of hylobatid evolution in what amounts to a so-called "ghost lineage" (Harrison 2016: 92).[6] Illumination is found in *Yuanmoupithecus* – a late Miocene (ca. 9.0-7.0 mya) fossil taxa from Yunnan in southern China. Harrison (2016: 100) concludes "that *Yuanmoupithecus* is the only known stem hylobatid and the only fossil representative of the clade that predates the Pleistocene". Unlike the hylobatids, the Asian great ape lineage (Hominidae: Ponginae) is relatively well documented and traces back to a Eurasian origin. The orangutan lineage preserves the evolutionary trajectory of larger body size (buffering both predation and energetic requirements) within a relatively narrow ecological niche, as described earlier in this section (see also Reichard et al. 2016). Meanwhile, the founders of the African ape and human clade likewise trace their evolutionary origins to the middle Miocene in Europe. Ultimately, the most parsimonious scenario is that in the early stages of the late Miocene, a few derived representatives from the diverse array of Eurasian hominoids return to the continent of Africa (Disotell & Stewart 1998). Their last remaining descendent species are the gorilla, chimpanzee, bonobo and humans.

In reviewing the evolutionary history of the hominoids, I have identified two salient insights. First, the dual challenges of an incomplete fossil record and the likelihood of widespread homoplasy within the evolutionary history of the hominoids, means that accurate reconstructions of precise ancestor–descendent relationships are exceedingly difficult to ascertain. Second, it is my view that in light of these challenges, it is more productive to draw inferences from the overall trends and trajectories across multiple lineages, than it is to place too much evolutionary import on the reconstruction of any one specific lineage of putative sister taxa. We've already seen the challenges that morphological trait mosaicism – "where more advanced features first appear in multiple species rather than aggregating in a single lineage over time" (McNulty 2010: 325) – presents for developing linear scenarios driven by singular selection pressures and/or singular evolutionary mechanisms. A focus on general trends rather than specific homologous (or indeed analogous) features helps to mitigate these challenges. Collectively these insights point towards principles I introduced in Chapter 1, those of *contingency* and *constrained totipotentiality*. In the interaction of antecedent states (see Chapter 1, Note 4) and the channelling of existing variation along pathways (which are partially shaped by the organisms themselves), we find a powerful, dialectic mode of analysis. Moving forward, as we dive deeper into the living hominoids and an emphasis on evolved patterns of behaviour, the further operationalisation of these concept will help us to tease apart and reconstruct emergent, social phenomena.

The living hominoids

In this section, I focus on the living apes, their diversity and socioecological characteristics. As previously described, the superfamily Hominoidea is a taxonomic category that comprises the families Hylobatidae (small apes, also known as the gibbons and siamang)[7] and Hominidae (great apes and humans)

Table 2.1 Taxonomy of the living members of the superfamily Hominoidea

Superfamily	Family	Subfamily	Tribe	Genus
Hominoidea				
	Hylobatidae (small apes)			*Hoolock* (3 species)
				Hylobates (9 species)
				Nomascus (7 species)
				Symphalangus (1 species)
	Hominidae (great apes and humans)			
		Ponginae		*Pongo* (3 species)
		Homininae		
			Gorillini	*Gorilla* (2 species)
			Panini	*Pan* (2 species)
			Hominini	*Homo* (1 species)

(Table 2.1). In addition to derived features of the postcranial skeleton (including a suite of adaptations related to the stability of forelimb joints through a range of suspensory and climbing motions), the living hominoids share an evolved emphasis on: (a) cognition and learning, underpinned by a particular neurological complexity, especially in the cortical areas of the brain; (b) extended life history phases, such as long periods of sub-adulthood and (c) the formation of complex social relationships (Larson 1998; MacLeod et al. 2003; Reichard 2009; Hartwig 2011). In summarising the available data on hominoid ecology and social behaviour, I hope to elucidate the commonalities and differences among taxa. For organisational purposes, I will address taxa primarily at the level of the genus (following Malone et al. 2012). However, this scheme should not lead invariably to assumptions of within-genus homogeneity with regard to feeding and socioecology. In some instances, I will describe species-level (or occasionally subspecies-level) variation. The aim of this overview is to form a baseline, hominoid trajectory of ecological and social system dynamics.

The Hylobatidae

The Hylobatidae is distinguished from the Hominidae by a suite of commonalities related to their highly mobile and energetically efficient mode of forelimb suspensory locomotion, as well as small body size, small group size and territorial behaviour (Jablonski & Chaplin 2009). The lineages of the four extant hylobatid genera experienced a rapid, and relatively recent, diversification at approximately 4–6 mya (Carbone et al. 2014). Today we recognise 20 hylobatid species, inclusive of the recent naming of the Gaoligong hoolock or "skywalker" gibbon (Anandam 2013; Roos 2016; Fan et al. 2017) (Table 2.2). Coupled with their acrobatic acumen and derived form of brachiation,[8] a markedly high

Table 2.2 Taxonomy of the family Hylobatidae

Genus (# of diploid chromosomes)	Species	Common Name
Hoolock (38)	hoolock	Western Hoolock gibbon
	leuconedys	Eastern Hoolock gibbon
	tianxing	Skywalker gibbon
Hylobates (44)	abbotti	Abbott's grey gibbon
	funereus	Eastern Bornean grey gibbon
	agilis	Agile gibbon
	albibaris	Bornean white-bearded gibbon
	klossii	Kloss's gibbon
	lar	White-handed gibbon
	moloch	Silvery gibbon
	muelleri	Müeller's gibbon
	pileatus	Pileated gibbon
Syndactylus (50)	symphalangus	Siamang
Nomascus (52)	hainanus	Hainan crested gibbon
	concolor	Western black-crested gibbon
	nasutus	Eastern black-crested gibbon
	leucogenys	Northern white-cheeked crested gibbon
	siki	Southern white-cheeked crested gibbon
	annamensis	Northern yellow-cheeked crested gibbon
	gabriellae	Southern yellow-cheeked crested gibbon

degree of genetic (e.g., discrete, lineage-specific diploid chromosome numbers, or 2n), pelage and vocal variation suggest that the Hylobatidae represents a unique phase of hominoid evolution (Figure 2.4). Additionally, though long considered to exhibit relatively uniform social structures, we now appreciate the range of hylobatid behavioural variation, both within and between the genera. Indeed, increased adaptability and ecological tolerance may underpin their "success" (in terms of species richness and continuous geographic distribution) relative to the large-bodied hominoids (Reichard et al. 2016). Let's briefly explore the four hylobatid genera.

The genus Hylobates

The genus *Hylobates* (2n = 44) is represented by nine species making it the most speciose of all the living ape genera. Species within the genus *Hylobates* are among the most well studied of all hylobatid taxa. Over a period of four decades, comparative data on the behavioural ecology of several *Hylobates* species have been published (e.g., Ellefson 1974; Raemaekers 1979; Kappeler 1984; Srikosamatara 1984; Leighton 1987; Palombit 1997; Brockelman et al. 1998; Bartlett 2009; Reichard et al. 2012). Unfortunately, for most species, key elements of the social systems, such as processes of pair formation/maintenance and intra- and inter-group behaviour, have received little attention.

Figure 2.4 The Javan silvery gibbon (*Hylobates moloch*), just one of 20 recognised species of hylobatid.

Photo credit: Ajat Surtaja.

However, the white-handed gibbon (*Hylobates lar*) is especially well known from two sites: Khao Yai National Park in Thailand (e.g., Brockelman et al. 1998; Reichard 2009; Reichard et al. 2012) and Ketambe Research Station in the Gunung Leuser National Park, Sumatra (Palombit 1992, 1994, 1996). Detailed observations of reproduction, natal dispersal, pair formation and group structure in the Khao Yai population, from 1978 to the present, represent the largest data set available with which to address questions about hylobatid social organisation. Specifically, the static model of highly territorial, monogamous pairs living in nuclear family groups can be assessed with data encompassing the life histories for individuals in several neighbouring social groups. A genetic relationship between adults and immatures in social groups is usually assumed in the nuclear family model. The findings of the many researchers at Khao Yai raise questions about these assumptions (Brockelman et al. 1998; Reichard 2009).

In a previous publication, Agustín Fuentes and I took a close look at some of the salient studies in an effort to destabilise the "nuclear family" assumption (Malone & Fuentes 2009). In this respect, the contributions of Warren Brockelman, Ulrich Reichard and colleagues represent an important record of long-term, population-level information. Data from a sample of

64 white-handed gibbon groups at Khao Yai reveal that 33% of the groups contained young estimated to be less than two years apart in age (Brockelman et al. 1998). Yet it is well documented the female hylobatids give birth, on average, once every three years. These data, in combination with relatively short dispersal distances, suggest that it is not uncommon for groups to contain young from more than one female, and that related individuals (e.g., siblings, cousins) may frequently reside in neighbouring territories. The potentially large number of non-nuclear families in this population stands in stark contrast to the conclusions of earlier researchers (summarised by Leighton 1987) as to the invariably monogamous social and reproductive behaviour of hylobatids (Fuentes 2000). Further challenges to the nuclear family model come from direct observation of partner turnover and extra-pair copulations (EPCs), reported to comprise up to 12% of all copulations (Reichard 1995, 2009). The implications of these observations for the nature of inter-group interactions are critically important. Given the potential inter-relatedness of gibbon groups, aggressive interactions with neighbours may in fact be detrimental to an individual's own inclusive fitness (Yi et al. 2020). Indeed, Bartlett (2003) reports that while a majority of inter-group encounters were considered agonistic, 20% consisted of vocal exchanges only, 6% were neutral and 17% included affiliation. Data such as these emphasise the importance of conceptualising the gibbon "group" to be intimately interconnected to "rich networks of genetic and social ties" at the community level (Whittaker & Lappan 2009: 5).

The genus Symphalangus

Siamang (*Symphalangus syndactylus*; $2n = 50$) are the sole monotypic genus within the family Hylobatidae. Siamang are relatively well known from Sumatra and the Malay Peninsula. The siamang is the largest of all hylobatid species, and is found in primary and secondary, lowland and montane forests up to 3800 metres above sea level. The larger bodied siamang (~10–11 kg) have long been considered the folivores of the family. However, Elder (2009) failed to discriminate between siamang and the small-bodied hylobatids (*Hylobates*, *Hoolock* and *Nomascus*) on several key dietary variables. The revelation here is that there is no significant relationship between body mass and folivory among the four hylobatid genera. Furthermore, general patterns of dietary and ecological variation, consistent with those found among species within the genus *Hylobates*, are evident between the insular (Sumatran) and mainland subspecies of siamang (Chivers 1974; Raemaekers 1979; MacKinnon & MacKinnon 1980; Palombit 1992; Lappan 2005). It is now clear that the folivorous-frugivorous siamang, like and other hylobatids (commonly referred to as ripe-fruit specialists), demonstrate a level of dietary breadth and ecological tolerance that was previously under-appreciated.

In the realm of social behaviour, studies of siamang have yielded insightful results. A relatively recent, long-term investigation of siamang groups at the Way Canguk Research Station in southern Sumatra has provided an insightful look into the social complexity of siamang populations (Lappan 2005, 2007). The occurrence of cohesive, greater-than-two-adult groups (multimale/unifemale composition) in four of the five focal groups, and the observation of poly-androus mating patterns in three of the four groups with more than a single adult male, provided an opportunity for the examination of male–infant, male–male and male–female relationships (Lappan 2007). Lappan's conclusions are relevant to my aims here, as they provide an opportunity to test long-held assumptions about the relationship between ecological factors (e.g., distribu-tion of resources) and social arrangements (e.g., pair bonds). In the multimale groups at Way Canguk, Lappan (2007) reports low rates of overall aggression, mutual tolerance among males, and variability in the strength and signalling of male–female bonds. Furthermore, Lappan's analyses of both genetic and behav-ioural data indicate that genetic relatedness cannot always be predicted from observable social relationships (Lappan 2005, 2007).

The genus Hoolock

Three species are now recognised in the genus *Hoolock* ($2n = 38$). The recently discovered Gaoligong or "skywalker" hoolock gibbon (*Hoolock tianxing*) joins the Western and Eastern hoolock gibbons (*H. hoolock* and *H. leuconedys*, respect-ively) as rare apes threatened with a heightened risk of extinction (Fan & Bartlett 2017; Fan et al. 2017). Hoolock gibbons inhabit rainforests and semi-deciduous forests in tropical and subtropical areas of India, Myanmar, China and Bangladesh. Large rivers play a significant role as barriers in the east–west dis-tribution of hoolock gibbons, and morphological and genetic variation within the genus follows suit. Very little is known about the behavioural ecology of this genus. Hoolock gibbons are the only hylobatids that range substantially outside of the tropics (Mootnick et al. 1987; Das & Biswas 2009). A predicted dietary response to a seasonal environment might include increased reliance on leaves as fruit availability decreases. However, Gittins and Tilson (1984) studying at a sub-tropical locale report a stable consumption of fruit throughout the year at a level comparable to hylobatid species of the genus *Hylobates*. Hoolock gibbons have been observed to occur in greater-than-two adult groups, including: two adult males and a female; two adult females and a male; and up to five adult males in a bachelor group in north-east India. Overall, four of the 34 hoolock groups on which studies have been published (12%) had a composition that runs counter to the traditional view of hylobatid social organisation (Tilson 1979; Siddiqi 1986; Choudhury 1990; Mukherjee et al. 1991–1992; Ahsan 1995).

The genus Nomascus

Understanding the basic range of diversity within the crested gibbon genus (*Nomascus*, $2n = 52$) is very much an ongoing affair, including the recent clas-sification of a seventh species (Thinh et al. 2010). As such, relatively few studies of the genus *Nomascus* have been conducted (e.g., Haimoff et al. 1987; Zhenhe et al. 1989; Bleisch & Chen 1991; Sheeran 1993; Fan et al. 2006; Geissmann 2007; Ruppell 2007; Hon et al. 2018). Species in this genus range in montane forests up to 2900 metres above sea level (asl). With the exception of the siamang, this represents the highest recorded altitude for a gibbon species (Bleisch & Chen 1991). Like some populations of hoolock gibbons, the crested gibbons can be found in drier, more seasonal habitats associated with higher latitudes. These seasonal, subtropical habitats differ from the tropical forests usually associated with hylobatids. The diet of the western black-crested gibbon (*Nomascus concolor*) has been reported to be more folivorous than those of other gibbons of similar size, and black-crested gibbons have been observed to come to the ground to forage on bamboo shoots. More recent studies have documented a degree of dietary selectivity in the eastern black-crested gibbon (*N. nasutus*) (Ma et al. 2017). One of the only studies of the most recently identified northern yellow-cheeked crested gibbon (*N. annamensis*) reports a diet consisting of just over 60% fruit (Hon et al. 2018).

With respect to crested gibbon social behaviour, early studies of group structure reported one-male/multifemale groups of up to four females, and average group sizes of 5.25 individuals (Haimoff et al. 1987). Overall, published reports indicate that approximately 25% of *Nomascus* spp. groups have greater than two adults, including observations of groups of up to 10 individuals including multiple females with dependent infants (Haimoff et al. 1986; Bleisch & Chen 1991; Lan & Sheeran 1995). Day range sizes are generally larger than those found for other gibbon species. Group size/composition was determined for seven groups. Groups were no larger than five individuals and well within the usual range for gibbons. Three groups consisted of a single male and multiple individuals who produced female vocalisations. The presence of multiple female singers in a single group is presented as evidence that these gibbons live in polygynous groups (Bleisch & Chen 1991). However, as immature individuals produce female songs in this species, this conclusion is tenuous. It is important to note that polygyny is one of many mating patterns that may exist within this particular social organisation. Similarly, while monogamy is often assumed, in reality there is a wide range of diversity in mating patterns in primates whose grouping patterns are predominated by two-adult pairs (Fuentes 2000, 2002).

In the previous section, I have attempted a(n) (admittedly) superficial synthesis of the published literature on the Hylobatidae. Let it be clear that our knowledge of this fascinating and diverse family remains incomplete. For example, direct observations of mating patterns, population-level genetic profiles and long-term behavioural studies of many hylobatid taxa are needed before interspecific comparisons can be fully realised. I do hope to have been successful in one modest task: the dispelling of the caricatured, inflexible "gibbon". Overall, the recognition of socioecological flexibility, in light of the highly specialised morphological characters that unify the small apes (and arguably constrain certain aspects of behaviour and ecology), remains of great interest and underscores the importance of a comparative, evolutionary approach (Lappan & Whittaker 2009; Reichard et al. 2016). We will return to the Hylobatidae in Chapter 3 to examine how our theoretical models of ape evolution must also "evolve" in light of this emergent picture of socioecological flexibility.

The Hominidae

Great apes and humans (Hominidae) share several morphological, physiological and social commonalities. Overall, slow progression through the life stages, combined with selection for cognition and learning, is a constant and consistent theme throughout the evolutionary trajectory of the Hominidae. Debates regarding the taxonomic relationships among the great apes and humans, and subsequently the use of certain taxonomic nomenclature, primarily centre around two alternative philosophies of classification (i.e., either prioritising ancestor/descendent relationships, or taking into account both descent and overall biological similarity) (Marks 2012). Here, following Harrison (2013), I emphasise genetic ancestry by delineating two subfamilies within the family Hominidae: (1) the Ponginae (orangutan); and (2) the Homininae (the African great apes and humans) (Table 2.3). What follows is a brief summary of the Hominidae, with an emphasis on social and ecological variation. I'll begin with the Asian great ape subfamily Ponginae (genus *Pongo*), and then proceed to review the Homininae (genera *Gorilla*, *Pan* and *Homo*).

The genus Pongo

Orangutans (Ponginae) diverged from the African great ape and human clade (Homininae) approximately 12–14 mya.[9] Once widely distributed (including populations in mainland and insular Southeast Asia, including Java), today orangutans persist only within fragmented forests on the islands of Sumatra and Borneo. The biogeographic and phylogenetic history of orangutans is quite complex as ancient patterns of sea-level fluctuations would have, at various times, exposed, connected and isolated the land masses that comprise the Sunda Shelf (Caldecott & McConkey 2005; Steiper 2006). Sumatran and Bornean orangutans are known as *P. abelii* and *P. pygmaeus*, respectively (Groves 2001).

Table 2.3 Taxonomy of the family Hominidae

Subfamily	Tribe	Genus	Species	Common Name
Ponginae		*Pongo*	*pygmaeus*	Bornean orangutan
			abelii	Sumatran orangutan
			tapanuliensis	Tapanuli orangutan
Homininae	Gorillini	*Gorilla*	*berengei*	Eastern gorilla
			gorilla	Western gorilla
	Panini	*Pan*	*troglodytes*	Chimpanzee
			paniscus	Bonobo
	Hominini	*Homo*	*sapiens*	Human

Additionally, some researchers recognise distinct lineages of Bornean orangutan at the subspecies, or even population-level (Warren et al. 2001), corresponding to various distribution patterns in Kalimantan (Indonesian Borneo) and the Malaysian states of Sabah and Sarawak. Recently, however, orangutan taxonomy and phylogeny have become much more interesting with the identification of a third distinct species, the Tapanuli orangutan (*P. tapanuliensis*) "that encompasses a geographically and genetically isolated population found in the Batang Toru area at the southernmost range limit of extant Sumatran orangutans, south of Lake Toba, Indonesia" (Nater et al. 2017: 3488). The strong case for the identification of a new species, based on morphological, genomic and behavioural evidence[10], as put forth by Alexander Nater and an international team of research collaborators, has also shed light on the evolutionary histories among the (now three) living orangutan species. The deepest evolutionary split (at ca. 3.4 mya) appears to be between Sumatran orangutan populations north (*P. abelii*) and south (*P. tapanuliensis*) of Lake Toba, although a signature of limited gene flow via dispersing males is evident until relatively recently (Nater et al. 2017). Phylogenetic analysis identified a much more recent split (ca. 700,000 years ago) between Bornean and Tapanuli orangutan. This addition of a new, critically endangered orangutan species presents several challenges to conservationists already struggling with emerging threats, limited resources and debates about preserving genetic purity. I will revisit these debates in Chapter 5.

As with the Hylobatidae, the backdrop of pronounced resource fluctuations in Southeast Asian forests is a cornerstone of grounding individual and group strategies in an ecological context. Southeast Asian rainforests are characterised by inter- and intra-annual variability in resource availability, driven in part by the interaction of climatic events (El Niño/Southern Oscillation) and mast fruiting by trees of the family Dipterocarpaceae (Whitmore 1984; Fleming et al. 1987; Ashton 1988; Curran et al. 1999). Fluctuations in the quantity and quality of available resources in Southeast Asian rainforests, including high variability in nutrient intake related to percentage of fruit in the diet, can substantially impact large-bodied, strictly arboreal mammals (Knott & Kahlenberg 2011). Indeed, orangutan are the largest, primarily arboreal mammal on the planet, and possess adaptations for an arboreal style of locomotion referred to as quadrumanous clambering (Figure 2.5). Orangutans occupy almost exclusively the high canopy of peat and freshwater swamps, extremely wet lowland forests and mountainous slopes up to an altitude of 1200 metres asl. Orangutans have an extraordinarily diverse diet, primarily consisting of fruit, but also include other plant parts (e.g., leaves, flowers, cambium), insects and vertebrates, on occasion in some locations (Russon et al. 2009). Females invest heavily in offspring, and only produce offspring every seven to nine years on average. Additionally, they display a high degree of sexual dimorphism in body mass (males can weight upwards of 75 kg). At sexual maturity, adult males can present as one of two different forms (bimaturism): (1) large dominant males with secondary sexual characteristics (flaring, fibrous cheek flanges, long hair on their arms and back, and extended throat sacs) that aggressively defend territories; or (2) unflanged,

Figure 2.5 A wild (non-rehabilitated) adult female Sumatran orangutan (*Pongo abelii*) and offspring at Bukit Lawang.

Photo credit: Lotte van den Hout.

more mobile individuals that also sire offspring on occasion (Maggioncalda et al. 2002; Ancrenaz et al. 2017).

The aforementioned morphological and physiological parameters, in combination with the ecology of their forested habitats, are central to explanations of orangutan social organisation. In addition to a semi-solitary existence within highly overlapping ranges, orangutan females (and immature offspring) form social groupings that consist of travel bands, feeding aggregations and consortships with individual males (Utami et al. 1997; Knott et al. 2008). Male ranges are large and overlap with both male and female ranges (Knott 1999; van Schaik 1999). Perhaps this is best described as dispersed sociality, or individual-based fission-fusion groupings. An important aspect of orangutan socioecology is the fact that in areas of high fruit density/low feeding competition (e.g., Suaq Balimbing, Sumatra) they radically alter their behavioural profiles to engage in much higher rates of affiliation. The flexibility inherent (and observed) in the dispersed sociality of orangutans is dramatic (especially if you include captive colonies). This increased sociality is associated with the innovation and diffusion of social traditions, including extra-somatic extractive foraging (van Schaik et al. 2003). To summarise, orangutan, while generally occurring in small groups or as individuals, do form larger aggregations at particularly resource-rich sites.[11] I, along with Agustín Fuentes and Frances White, have argued that this represents a capacity for socioecological flexibility of critical (past and ongoing) evolutionary importance (Malone et al. 2012). Next, I will turn my attention towards the Homininae, or the African great apes and humans. Here, similarly, I'll describe species as occurring primarily in some variant of multimale/multifemale groups that exhibit variable patterns of cohesion, albeit more reliant on terrestrial lifestyles. I hope to demonstrate a "continuity of capacity" that connects all the apes, and will revisit the theoretical ramifications of an emergent, variable sociality in Chapter 3.

Figure 2.6 Mountain gorilla (*Gorilla beringei beringei*) in Rwanda.
Photo credit: Alison Wade.

The genus Gorilla

The genus *Gorilla* is divided into eastern and western species (Vigilant & Bradley 2004). Furthermore, two subspecies of eastern gorillas (*Gorilla beringei*) are recognised: eastern lowland gorillas (*G. b. graueri*) occupy eastern forests of the Democratic Republic of the Congo (DRC), while mountain gorillas (*G. b. beringei*) inhabit mountainous areas of Uganda, Rwanda and the DRC (Figure 2.6). Likewise, two subspecies of western gorilla (*G. gorilla*) are recognised with western lowland gorillas (*G. g. gorilla*) relatively widely distributed across Gabon, Cameroon, Equatorial Guinea, the Central African Republic, Angola and the DRC, whereas Cross River gorillas (*G. g. diehli*) are severely restricted to a limited expanse of protected and unprotected forests in Cameroon and Nigeria (Cooksey & Morgan 2017; Wade 2020). All taxa are very large and exhibit a strong degree of sexual dimorphism in both body mass and skeletal anatomy. It is relevant to note that large body size likely acts to exert a wide array of specific constraints on gorillas, relative to the other hominoids. The nutritional requirements of large body size, in combination with a relatively generalised digestive anatomy and physiology, account for the primarily frugivorous-folivorous diet of gorillas (Harcourt & Stewart 2007). Western and eastern gorillas differ by degrees of frugivory and arboreal habits, both of which are markedly more common in the former (Dunn et al. 2014; Lodwick & Salmi 2019).

Gorilla socioecology is largely informed by numerous studies of the mountain-dwelling populations of eastern gorillas (e.g., Doran & McNeilage 1998; Grueter et al. 2013), although recent research is providing new insights into western gorilla sociality, diet and ranging (e.g., Remis et al. 2001; Cipolletta 2004; Robbins et al. 2004; Rogers et al. 2004; Cooksey et al. 2020). Gorillas form stable, cohesive groups of between 8 and 10 individuals, although much larger groups (>20 individuals, up to 40+) have been observed (Tutin 1996;

Yamagiwa et al. 2003; Robbins 2011). Patterns of male and female philopatry, dispersal and inter-group transfer are relatively complex in this genus. Female primary and secondary dispersal is the norm, while males may remain in their natal groups post-maturity, or disperse (either alone or with a subset of females) to join form their own group, join an all-male group or become temporarily solitary (Robbins et al. 2004; Caldecott & Ferriss 2005; Robbins 2011). Additionally, Goldsmith (1996) observed multimale/multifemale groups that fissioned into subgroups during times of low resource availability. Accounting for all of this complexity, gorillas are best characterised as exhibiting a dynamic sociality comprising a continuum of inter-individual associations from cohesive one-male or multimale/multifemale groups to dispersed one-male or multimale subgroups (Doran & McNeilage 1998; Fuentes 2000; Forcina et al. 2019).

Within-group social relationships among gorillas are most influenced by the aforementioned patterns of philopatry and dispersal, as well as the distribution of resources and the potential competition among females for male services (e.g., protection against extra-group males) (Watts 2001; Robbins 2011). Abundant and evenly distributed resources (i.e., reduced levels of contest competition) are most likely the underpinnings of weak social bonds and/or dominance relationships among female gorillas (Harcourt 1979; Watts 2001), although Robbins et al. (2005) document long-term stability in the female–female dominance relationships that do exist. In contrast, male–female relationships are the core of intra-group sociality and are characterised by high levels of affiliation (especially with respect to proximity maintenance), as well as the sustained, low-level agonistic behaviour of males directed towards females (Watts 1992; Sicotte 1994; Robbins 2003, 2011). The nature of inter-group encounters ranges from aggressive to peaceful, with the degree of relatedness among males and number of potential migrant females accounting for much of the variance (Sicotte 1993; Bradley et al. 2004). Additional insight is provided from a recent long-term study of inter-group encounters among western lowland gorillas at two, ecologically similar sites in the Republic of Congo (Cooksey et al. 2020). Kristena Cooksey and colleagues determined that social factors, such as extended, extra-group social networks, were more important than ecological factors in predicting the outcomes of inter-group encounters. Networks of familiar males were associated with both the likelihood of engaging in inter-group encounters and also the nature of the encounters, with familiarity leading to increased social tolerance and even affiliation between groups (Cooksey et al. 2020). In sum, despite certain constraining factors related to body size and the concomitant nutritional requirements within the genus *Gorilla*, the observed ranges of behavioural variation in social organisation (i.e., group composition and cohesion) and social structure (i.e., the nature of relationships within and between the sexes) support my argument that an underpinning, adaptive plasticity is a core component of this lineage's evolutionary trajectory.

The genus Pan

Humankind's closest living lineage (tribe Panini: genus *Pan*) comprises two species, the robust chimpanzee (*Pan troglodytes*) and the gracile chimpanzee, or bonobo (*P. paniscus*) (Figure 2.7). Chimpanzees are now generally recognised as having four subspecies: the eastern chimpanzee (*P. t. schweinfurthii*), the central chimpanzee (*P. t. troglodytes*), the western chimpanzees (*P. t. verus*) and the Nigerian-Cameroon chimpanzee (*P. t. ellioti*). These species and subspecies designations within the genus *Pan* are well supported on the basis of geographical distribution, genetic variation and morphological distinctions (Pilbrow

Figure 2.7 Two species within the genus *Pan*. Above: an adult male bonobo (*Pan paniscus*) "Larry" at the Iyema study site, D.R.C. Below: an adult female chimpanzee (*Pan troglodytes*) "Ekisigi" with her dependent off spring "Euro" in Kibale National Park, Uganda.

Photo credits: Amy Cobden and Alain Houle.

2006; Stumpf 2011), and reflective of a complex demographic history (Fischer et al. 2004). Extensive datasets on chimpanzees have come from long-term studies, most notably Gombe (e.g., Goodall 1986) and Mahale (e.g., Nishida 1990). Chimpanzees live in a wide diversity of habitats from highly productive lowland rain forests, to montane and bamboo forests, and extending to dry savanna or woodland mosaic habitats. The behavioural and ecological flexibility of chimpanzees, combined with the historical oscillations of broadly suitable habitats in the Afrotropics, informs to the patterning of genetic and behavioural diversity both within and between chimpanzee subspecies (Barratt et al. 2020). Bonobos, in contrast, are confined to the Congo Basin, but are also found in a diversity of habitats, from poly-specific evergreen rainforest with little human disturbance, to mixed primary and secondary forest in close association with human populations. Studies of both bonobos and chimpanzees vary greatly in length, degree of habituation and history of provisioning, as well as the direct and indirect effects from local human populations, all of which may contribute to demographic differences between studies.

Chimpanzees and bonobos at all sites consume a highly frugivorous diet, supplemented with leaves, flowers, pith, insects and meat (Nishida & Hiraiwa-Hasegawa 1987; White 1996). Seasonality, and the corresponding variation in fruit abundance, is less marked for bonobo populations, relative to comparable chimpanzee sites (White 1998; Doran et al. 2002). In general, highly productive sites are associated with more fruit consumption and more diverse diets (Stumpf 2011). There is also considerable variation among study populations in the amount of meat eating, hunting and tool-use to obtain insects and other food items. Both chimpanzees (Nishida et al. 1979; Goodall 1986; Boesch 1994; Watts & Mitani 2002) and bonobos (Badrian & Malenky 1984; White 1994; Hohmann & Fruth 2008) take animal prey, although there is considerable variation in hunting or capture behaviour, as well as prey species taken among study populations. Our knowledge of chimpanzee hunting practices is currently expanding to include understudied environments such as in the arid (woodland–savannah) landscapes of south-eastern Sénégal. Recent work by Professor Jill Pruetz and colleagues has documented a diverse range of mammalian prey (including multiple primate species) taken by western chimpanzees (Pruetz et al. 2015). Notably, galagos (small, nocturnal primate species) represent a high percentage of captured prey by the chimpanzees at the site of Fongoli, including by way of tool-assisted hunting (especially by females). Indeed, the manufacture and/or use of tools to obtain food is more commonly observed in chimpanzee studies (McGrew 1992), but is also known for bonobos (Badrian et al. 1981; White et al. 2008). Observed, species-level differences likely represent behavioural responses to habitat productivity.

Both bonobos and chimpanzees can be broadly defined as showing fission-fusion social systems with communities of multiple males and females with male philopatry. Chimpanzees are male-bonded with strong social ties among related males. Chimpanzee communities can vary from small stable units with only one male as at Bossou (albeit under uncommon conditions) (Sugiyama

& Koman 1979; Sugiyama 2004), to large units of 150–200 individuals (Watts 2008; Langergraber et al. 2017). Bonobo communities in contrast show strong relationships among females as well as between males and females (Furuichi 1987; White 1989). As in chimpanzees, bonobo communities also vary from those with smaller membership numbers (referred to as "splinter groups") centred around immigrating females that persist for between two and three years (White 1996) to larger long-term, stable communities of up to 50–120 individuals (Furuichi 1989; White 1996). As to be expected with the wide range of ecological conditions both among sites as well as seasonally within sites, party composition varies greatly and can often be correlated to food availability and/or the distribution of females (especially during periods of sexual attractivity, proceptivity and receptivity). In general, chimpanzee sociality reflects heightened male–male affiliation among both related and non-related males with typically low, but varying amounts, of female cohesion. Bonobo communities show strong relationships among (unrelated) females as well as between males and females (White 1986; Wrangham 1986; Morin et al. 1994; Hohmann et al. 1999; Doran et al. 2002; Lehmann & Boesch 2008). The greater female affiliation in bonobos may be possible through reduced feeding competition and the consumption of terrestrial herbaceous vegetation (Wrangham 1986), or the presence of larger feeding patches associated with the occupation of more productive habitats (Malenky 1990; White 1986; White & Wrangham 1988). Overall, the rates and intensity of aggresive interactions (both within- and between-communities) are reduced, though not absent, in bonobos as compared to chimpanzees (Sommer et al. 2011; Stumpf 2017; Lucchesi et al. 2020). Social or "cultural" traditions, boundary patrols, tool creation/use (more commonly observed in chimpanzees but also known for bonobos) and the varied nature of intersexual relationships, set within complex fission-fusion communities, demonstrate that for these great apes, most ecological challenges are tackled behaviourally and with substantial flexibility. We will explore ways to model these complex dynamics in Chapter 3.

The genus Homo

With respect to our own lineage (Tribe Hominini: genus *Homo*), a divergence at approximately 7–9 mya is marked by a shift towards adaptations that facilitate bipedal locomotion. This derived form of locomotion precedes the expansion of relative brain size. Simultaneously, the evolution of a distinct "childhood" phase, in association with extra-parental care, permits an increased reproductive capacity by shortening inter-birth intervals (Bogin 1999). In essence, our own evolutionary lineage (i.e., hominins) "severed the habitat bias", or constraint of relatively stable and productive forests characteristic of hominoid evolution since the Miocene, to expand into diverse environments and a (nearly) planet-wide distribution (Potts 2004: 209). In contemporary humans, we see the emergence of temporally and spatially complex alliances, inter- and intragroup social negotiations that involve symbolic, somatic and material components, and a

Figure 2.8 The author and his daughter negotiating symbolic, somatic and material terrain in Yogyakarta, Indonesia.

Photo credit: Meredith Malone.

level of social and material reciprocity not seen in other lineages (Figure 2.8). This ability to negotiate increasingly complex and information-rich social and ecological networks has become the primary avenue for achieving social and reproductive success (Sussman & Cloninger 2011). Contemporary humans, constructing social relationships in response to a myriad of interacting social, economic, political and ecological factors, exhibit an unparalleled range of complexity and flexibility in social organisation (Kendal et al. 2011; Fuentes 2017). It is reasonable to attribute such clearly manifested behavioural patterns to a ratcheting up of niche construction from a robust baseline of behavioural flexibility shared by all members of the family Hominidae (Malone et al. 2012).

From the Miocene to the margins

Thus far my focus has been, and will by and large remain, on the superfamily Hominoidea. However, many of the evolutionary trajectories described herein stem from basic, primate-wide trends (e.g., large investment in a relatively few offspring, large brains with enhanced neocortical development and complex sociality as an adaptive niche). Even general mammalian systems that we share in common with many co-occupants of our human niche (e.g., dolphin and dog, koala bear and hog[12]) allow us to identify with the behavioural expression to select other species. As fellow anthropoid primates and hominoids, our ability to observe these capacities within our fellow apes is enhanced by additional neuro-cognitive capacities and architecture.[13] As such, an understanding of social and ecological variation among the apes can provide a comparative evolutionary perspective into the socioecological trajectory of the human lineage. Furthermore, the phylogenetic proximity between apes and ourselves can produce evocative connections and compelling narratives. In this sense, our scientific and cultural relationships with the apes represent important waypoints in the navigation of humankind's relationship with the natural world and with each other.

In the present chapter, I have detailed the inter-relationships between eco-logical contexts and the evolutionary lineages of hominoids. Both of the two families (Hominidae and Hylobatidae) exhibit complex and context-dependent behaviour. Extant members of the family Hominidae (great apes and humans) are clearly characterised by varied patterns of complex social relationships that extend beyond biological kin, beyond reproductive relationships and beyond the immediate social group. The potential for the innovation, spread and inheritance of behavioural patterns and social traditions is much higher in the Hominidae than in other anthropoid primates. Richard Potts (1998, 2004) argues convincingly that these capacities evolved in response to the combin-ation of dietary/metabolic constraints and increasing seasonality/habitat vari-ability. For the small apes (Hylobatidae), although brain size is below that of the other apes, key aspects of brain architecture and proportionality are distinctively *hominoid*. For example, MacLeod et al. (2003) examined the structure and cir-cuitry of the cerebellum – areas of the brain integral for planning and executing complex movements and procedural learning – were shared among *all* apes and humans (to the exclusion of monkeys). Ulrich Reichard reinforces this idea of a *hominoid grade shift*, concluding that, based on evidence of sociosexual flexi-bility, hylobatids share with the great apes "a basic cognitive capacity for solving nonsocial problems with social solutions" (2009: 371). Finally, although many facets of hylobatid communicative and cognitive capacities remain unknown, promising work to develop multi-modal approaches and methodological enhancements are underway (Liebal et al. 2013; Liebal 2016).

It is important that we take (at least) two important lessons forward with us into the following chapter: (1) the influential role that historical constraint has played in shaping the socioecology of ape lineages; and (2) with respect to social system variation, we take seriously the interplay between behaviourally flexible individuals and emergent, population-level dynamics. Population-level process encompasses a host of factors including environmental conditions, historical constraints, density dependent effects and stochastic processes. This means that, at any given time, researchers should expect there to be an underlying tension between observing "behavioural equilibrium" and documenting optimal, adaptive trajectories (Oyama et al. 2001). In practice, primatologists tend to theorise the latter prior to setting-out and recording the former. Reflecting back on Walter Hartwig's quote at the onset of this chapter, it is clear to see the centrality of *both* constraint *and* behavioural potential in the patterns of "descent with modification" that are characteristic of hominoid evolution. The dialectical perspective I am advocating for gives primacy to these attributes. To better capture the complex interplay of individual, group and population dynamics, we'll need to incorporate these attributes into an enhanced, theoret-ical toolkit. In the following chapter, I aim to demonstrate the utility of these enhancements.

Finally, all of the ape species are threatened with a heightened risk of extinction, primarily due to humankind's increasingly acute alteration and/or destruction of ape habitats (Hockings et al. 2015; Estrada et al. 2017, 2020). Simply put, a small array of resilient and evolutionarily constrained apes cling to a fragile existence while one ex-ape (us) alters the planet to an unprecedented degree.[14] How we theorise and study apes has implications for how we conserve them. While the great apes receive a lot of funding and attention from researchers and conservationists, the hylobatids, historically suffering under the moniker of "lesser apes", have received less attention from the global research and conservation communities (Whittaker & Lappan 2009; Fan & Bartlett 2017). We have an urgent need to raise awareness of the threats facing all of the apes, and a fast-closing window of opportunity within which to act. In later chapters, I will be exploring aspects of the conservation crisis in further detail. Many questions pertaining to the relationship between research and conservation still remain. How can we prevent the loss of any ape species? Are we able to include evolutionary analyses in our primarily conservation-driven research agendas? Can our evolutionary models account for the behavioural responses by apes to the myriad ways that humans impact their populations? As failing to account for these dynamics can "risk scientific invalidity and conservation irrelevance" (Behie et al. 2019: 6), the incorporation of a critical, dialectical approach to primate research and conservation is justified.

Notes

1 The gibbons' distinctive characteristics, such as long limbs, fluid movements and intricate song bouts, often expressed as coordinated duets by pair-mates, underpin their revered status (Malone et al. 2014). For additional treatment of how apes have historically captured the collective attention of people the world over, see the proceedings (1995, edited by Raymond Corbey and Bert Theunissen) of the 1993 symposium titled *Ape, Man, Apeman: Changing Views since 1600*.

2 While not diminishing the overall finding that hominoid diversity declines in the late-Miocene period, it is always possible that new discoveries will alter our overall picture of species richness in the fossil record. Importantly, Harrison (2010: 464) alludes to: "tantalising evidence from the later Miocene of East Africa demonstrates that hominoids were quite diverse during this period, although few species have been formally named because of the paucity of the material".

3 Harrison (2002, 2010) and others (e.g., Ward & Duren 2002) recognise up to three non-cercopithecoid superfamilies, adding Dendropithecoidea and recognising Proconsuloidea as a distinct superfamily. In contrast, Fleagle (1999) groups the genera *Dendropithecus, Proconsul* and others all within a single family (Proconsulidae) within a singular superfamily (Hominoidea). While not discounting the value of such nuanced taxonomic debates and the quest to ascertain precise ancestor–descendent relationships, for our purposes here, a more general emphasis on evolutionary trends will suffice. Capturing this sentiment, Hunt (2016: 666), citing Harrison's earlier work (1982, 1987), states: "While some undiscovered species may yet prove to be the true ancestor of apes, the proconsuloids are a stem catarrhine

very close to the divergence of apes and monkeys and are therefore likely similar in morphology to the earliest apes, even if not directly ancestral".

4 Fleagle (1999: 454) describes the diagnostic features of proconsulid molars as such:

> the upper molars distinguish proconsulids from earlier propliopithecids and are characterised by their quadrate shape with a relatively larger hypocone, a pronounced, often beaded lingual cingulum, and some details of the conules (Kay, 1977). The lower molars have a broad talonid basin surrounded by five prism-like cusps, including a large hypoconulid.

This distinctive "Y-5 pattern" on the lower molars is the primitive condition for the Hominoidea and an extremely useful diagnostic tool in the identification of hominoids in the fossil record.

5 See McNulty (2010: 324) for a discussion of "stem-catarrhines", "stem-hominoids" (related, but not uniquely linked, to modern ape lineages) and stem taxa of the early Miocene (STEM).

6 Previously, some candidate taxa have been put forward (e.g., *Laccopithecus* from the late Miocene site of Lufeng, China) (Wu & Pan 1984), but subsequent analysis reveals this form to be a late-surviving pliopithecid (a distantly related, stem-catarrhine) (Harrison 2016).

7 Following Reichard et al. (2016: 4), "the terms 'hylobatids', 'small Asian apes', 'small apes', and 'gibbons and siamang' are used synonymously in this chapter and volume to describe the family Hylobatidae (Gray 1870)". However, departing from Reichard et al. (2016), I will not use the term "gibbons" to refer exclusively to the species within the genus "Hylobates", but instead will use clarifying language (e.g., hoolock gibbons) when referring exclusively to species within any of the four genera within the Hylobatidae.

8 As habitually (and often ricochetally) brachiating, small-bodied hominoids, the hylobatids occupy a unique phylogenetic position with respect to ecology and reproductive biology. Several energy-economising aspects of hylobatid locomotion and postural biomechanics have been identified, including: access to resources near the outer periphery of trees, the alternating transformation of potential energy into kinetic energy (the mathematical pendulum model) and increasing horizontal velocity via relatively long forelimbs (Preuschoft & Demes 1984).

9 Some support exists for both early (19–24 mya) (Easteal & Herbert 1997) and late (8–9 mya) (Arnason et al. 1996) divergence scenarios.

10 In addition to morphometric and genomic analyses, Nater et al. (2017) report both new observations and previously published descriptions of male "long call" vocalisations as behavioural evidence in support of the classification of extant orangutans into three species. In this comparison, 15 long call variables were analysed. The male long call of the Tapanuli orangutan (*P. tapanuliensis*) differs from *P. abelii* by having "a higher maximum frequency range of the roar pulse type (>800 versus <747 Hz) with a higher 'shape' (>952 versus <934 Hz/s)" (Nater et al. 2017: 3489). As distinct from *P. pygmaeus*, the Tapanuli orangutan male long call "has a longer duration (>111 versus <90 s) with a greater number of pulses (>52 versus <45 pulses), and is delivered at a greater rate (>0.82 versus <0.79 pulses per 20 s)" (Nater et al. 2017: 3489).

11 See also Roth et al. (2020) for between-site and between-species comparisons for orangutan species in Sumatra.

12 Extracted song lyrics from *Mammals* (1992), written and performed by They Might Be Giants.
13 The presence and activation of both von Economo and mirror neurons facilitate a wider range of potential behavioural response without requiring the selection of specific behavioural patterns. See Allman et al. (2010) and Ramanchandran (2011).
14 Given the intensity and scope of humankind's impact and influence on ecological processes, the *Anthropocene* is a proposed formal unit of geological time whereby humankind's "signature" of alterations to basic biogeochemical processes is written into the earth's strata (Crutzen & Stoermer 2000; Steffen et al. 2011; Ellis 2015).

References

Ahsan F. 1995. Fighting between two females for a male in the hoolock gibbon. *International Journal of Primatology* 16: 731–737.

Allman JM, Tetreault NA, Hakeem AY, Manaye KF, Semendeferi K, Erwin JM, Park S, Goubert V, Hof PR. 2010. The von Economo neurons in the frontoinsular and anterior cingulate cortex in great apes and humans. *Brain Structure and Function* 214: 495–517.

Anandam MV, Groves CP, Molur S, Rawson BM, Richardson MC, Roos C, Whittaker DJ. 2013. Species accounts of Hylobatidae. In Mittermeier RA, Rylands AB, Wilson DE. (Eds.), *Handbook of the Mammals of the World: Primates* (Vol. 3, pp. 778–791). Barcelona: Lynx Edicions.

Ancrenaz M, Oram F, Lackman I. 2017. Orangutan (*Pongo*). In *The International Encyclopedia of Primatology* (pp. 1–2). Hoboken, NJ: John Wiley & Sons.

Andrews P. 1983. The natural history of *Sivapithecus*. In Ciochon RL, Corruccini RS. (Eds.), *New Interpretations of Ape and Human Ancestry* (pp. 441–463). Boston, MA: Springer.

Andrews P. 1986. Fossil evidence on human origins and dispersal. In *Cold Spring Harbor Symposia on Quantitative Biology* (Vol. 51, pp. 419–428). New York, NY: Cold Spring Harbor Laboratory Press.

Andrews P. 2020. Last common ancestor of apes and humans: Morphology and environment. *Folia Primatologica* 91(2): 122–148.

Andrews P, Martin L. 1987. Cladistic relationships of extant and fossil hominoids. *Journal of Human Evolution* 16(1): 101–118.

Arnason U, Gullberg A, Janke A, Xu X. 1996. Pattern and timing of evolutionary divergences among hominoids based on analyses of complete mtDNAs. *Journal of Molecular Evolution* 43(6): 650–661.

Ashton PS. 1988. Dipterocarp biology as a window to the understanding of tropical forest structure. *Annual Review of Ecology and Systematics* 19(1): 347–370.

Badrian NL, Badrian AJ, Susman RL. 1981. Preliminary observations on the feeding behaviour of *Pan paniscus* in the Lomako Forest of Central Zaire. *Primates* 22: 173–181.

Badrian NL, Malenky RK. 1984. Feeding ecology of *Pan paniscus* in the Lomako Forest, Zaire. In Susman RL. (Ed.), *The Pygmy Chimpanzee: Evolutionary Biology and Behaviour* (pp. 275–299). New York, NY: Plenum Press.

Barratt CD, Lester JD, Gratton P, Onstein RE, Kalan AK, McCarthy MS, Bocksberger G, White LC, Vigilant L, Dieguez P, Abdulai B, et al. 2020. Late Quaternary habitat suitability models for chimpanzees (*Pan troglodytes*) since the Last Interglacial (120,000 BP). *bioRxiv*.

Bartlett TQ. 2003. Intragroup and intergroup social interactions in white-handed gibbons. *International Journal of Primatology* 24(2): 239–259.

Bartlett TQ. 2009. Seasonal home range use and defendability in white-handed gibbons (*Hylobates lar*) in Khao Yai National Park. In Lappan S, Whittaker DJ. (Eds.), *The Gibbons: New Perspectives on Small Ape Socioecology and Population Biology* (pp. 265–275). New York, NY: Springer.

Behie AM, Teichroeb JA, Malone N. 2019. Changing priorities for conservation and research in the Anthropocene. In Behie AM, Teichroeb JA, Malone N. (Eds.), *Primate Research and Conservation in the Anthropocene*. Cambridge: Cambridge University Press.

Begun DR. 2007. Fossil record of Miocene hominoids. In Henke W, Tattersall I. (Eds.), *Handbook of Palaeoanthropology, Vol. 2: Primate Evolution and Human Origins* (pp. 921–977). Berlin: Springer.

Begun DR. 2017. Evolution of the Hominoidea. *The International Encyclopedia of Primatology* 1: 1–4.

Begun DR, Nargolwalla MC, Kordos L. 2012. European Miocene hominids and the origin of the African ape and human clade. *Evolutionary Anthropology: Issues, News, and Reviews* 21(1): 10–23.

Bleisch WV, Chen N. 1991. Ecology and behaviour of wild black-crested gibbons (*Hylobates concolor*) in China with a reconsideration of evidence for polygyny. *Primates* 32: 539–548.

Boesch C. 1994. Cooperative hunting in wild chimpanzees. *Animal Behaviour* 48: 653–667.

Bogin B. 1999. *Patterns of Human Growth*. Cambridge: Cambridge University Press.

Bradley BJ, Doran-Sheehy DM, Lukas D, Boesch C, Vigilant L. 2004. Dispersed male networks in western gorillas. *Current Biology* 14: 510–514.

Brockelman WY, Reichard U, Treesucon U, Raemaekers JJ. 1998. Dispersal, pair formation and social structure in gibbons (*Hylobates lar*). *Behavioural Ecology and Sociobiology* 42: 329–339.

Caldecott J, Ferriss S. 2005. Gorilla overview. In Caldecott J, Miles L. (Eds.), *World Atlas of Great Apes and their Conservation* (pp. 97–103). Berkeley: University of California Press.

Caldecott J, McConkey K. 2005. Orangutan overview. In Caldecott J, Miles L. (Eds.), *World Atlas of Great Apes and their Conservation* (pp. 153–159). Berkeley: University of California Press.

Carbone L, Harris RA, Gnerre S, Veeramah KR, Lorente-Galdos B, Huddleston J, Meyer TJ, Herrero J, Roos C, Aken B, Anaclerio R, et al. 2014. Gibbon genome and the fast karyotype evolution of small apes. *Nature* 513(7517): 195–201.

Chatterjee HJ. 2016. The role of historical and fossil records in predicting changes in the spatial distribution of hylobatids. In Reichard U, Hirai H, Barelli C. (Eds.), *Evolution of Gibbons and Siamang* (pp. 43–54). Developments in Primatology: Progress and Prospects. New York, NY: Springer.

Chivers DJ. 1974. The siamang in Malaya: A field study of a primate in tropical rain forest. *Contributions to Primatology* 4: 1–335.

Choudhury A. 1990. Population dynamics of Hoolock (*Hylobates hoolock*) in Assam, India. *American Journal of Primatology* 20: 37–41.

Cipolletta C. 2004. Effects of group dynamics and diet on the ranging patterns of a western gorilla group (*Gorilla gorilla gorilla*) at Bai Hokou, Central African Republic. *American Journal of Primatology* 64: 193–206.

Cooksey KE, Morgan DB. 2017. Gorilla (*Gorilla*). *The International Encyclopedia of Primatology* 1: 1–5.

Cooksey KE, Sanz C, Ebombi TF, Massamba JM, Teberd P, Magema E, Abea G, Peralejo JSO, Kienast I, Stephens C, Morgan D. 2020. Socioecological factors influencing intergroup encounters in western lowland gorillas (*Gorilla gorilla gorilla*). *International Journal of Primatology*. DOI:10.1007/s10764-020-00147-6.

Corbey R. 2005. *The Metaphysics of Apes: Negotiating the Animal-Human Boundary.* Cambridge: Cambridge University Press.

Corbey R, Theunissen B. 1995. *Ape, man, apeman: Changing views since 1600.* Evaluative proceedings of the symposium Ape, man, apeman: Changing views since 1600, Leiden, the Netherlands, June 28–July 1, 1993. Leiden: Department of Prehistory, Leiden University.

Crutzen PJ, Stoermer EF. 2000. The Anthropocene. *Global Change Newsletter* 41: 17–18.

Curran LM, Caniago I, Paoli GD, Astiani D, Kusneti M, Leighton M, Nirarita CE, Haeruman H. 1999. Impact of El Nino and logging on canopy tree recruitment in Borneo. *Science* 286: 2184–2188.

Das J, Biswas J. 2009. Review of western hoolock gibbon (*Hoolock hoolock*) and eastern gibbon (*Hoolock leuconedys*) diversity with a presence of another new distinct variety of gibbon in Arunachal Pradesh, India. *American Journal of Primatology* 71(Suppl. 1): 72.

Dean D, Delson E. 1992. Second gorilla or third chimp? *Nature* 359: 676–677.

Deane AS, Begun DR. 2008. Broken fingers: Retesting locomotor hypotheses for fossil hominoids using fragmentary proximal phalange and high-resolution polynomial curve fitting (HR-PCF). *Journal of Human Evolution* 55: 691–701.

Doran DM, Jungers WL, Sugiyama Y, Fleagle JG, Heesy CP. 2002. Multivariate and phylogenetic approaches to understanding chimpanzee and bonobo behavioural diversity. In Boesch C, Hohmann G, Marchant L. (Eds.), *Behavioural Diversity in Chimpanzees and Bonobos* (pp. 14–34). New York, NY: Cambridge University Press.

Doran DM, McNeilage A. 1998. Gorilla ecology and behaviour. *Evolutionary Anthropology* 6: 120–131.

Dunn RH, Tocheri MW, Orr CM, Jungers WL. 2014. Ecological divergence and talar morphology in gorillas. *American Journal of Physical Anthropology* 153(4): 526–541.

Easteal S, Herbert G. 1997. Molecular evidence from the nuclear genome for the time frame of human evolution. *Journal of Molecular Evolution* 44: S121–S132.

Elder AA. 2009. Hylobatid diets revisited: The importance of body mass, fruit availability, and interspecific competition. In Lappan S, Whittaker DJ. (Eds.), *The Gibbons: New Perspectives on Small Ape Socioecology and Population Biology* (pp. 133–159). New York, NY: Springer.

Ellefson JO. 1974. *Natural History of White-Handed Gibbons in the Malayan Peninsula.* PhD Dissertation, University of California, Berkeley.

Ellis EC. 2015. Ecology in an anthropogenic biosphere. *Ecological Monographs* 85(3): 287–331.

Estrada A, Garber PA, Chaudhary A. 2020. Current and future trends in socio-economic, demographic and governance factors affecting global primate conservation. *PeerJ* 8: e9816.

Estrada A, Garber PA, Rylands AB, Roos C, Fernandez-Duque E, Di Fiore A, … & Rovero F. 2017. Impending extinction crisis of the world's primates: Why primates matter. *Science Advances* 3(1): e1600946.

Fan P, Bartlett TQ. 2017. Overlooked small apes need more attention!. *American Journal of Primatology* 79(6): e22658.

Fan PF, He K, Chen X, Ortiz A, Zhang B, Zhao C, ... & Groves C. 2017. Description of a new species of hoolock gibbon (Primates: Hylobatidae) based on integrative taxonomy. *American Journal of Primatology* 79(5): e22631.

Fan P, Jiang X, Liu C, Luo W. 2006. Polygynous mating system and behavioural reason of black crested gibbon (*Nomascus concolor jingdongensis*) at Dazhaizi, Mt. Wuliang, Yunnan, China. *Zoological Research* 27: 216–220.

Fischer A, Wiebe V, Pääbo S, Przeworski M. 2004. Evidence for a complex demographic history of chimpanzees. *Molecular Biology and Evolution* 21(5): 799–808.

Fleagle JG. 1999. *Primate Adaptation and Evolution*, 2nd ed. San Diego, CA: Academic Press.

Fleming TH, Breitwisch R, Whitesides GH. 1987. Patterns of tropical vertebrate frugivore diversity. *Annual Review of Ecology and Systematics* 18(1): 91–109.

Forcina G, Vallet D, Le Gouar PJ, Bernardo-Madrid R, Illera G, Molina-Vacas G, ... & Bermejo M. 2019. From groups to communities in western lowland gorillas. *Proceedings of the Royal Society B* 286(1896): 20182019.

Frost SR. 2017. Evolution of the Cercopithecidae. *The International Encyclopedia of Primatology*: 1–3.

Fuentes A. 2000. Hylobatid communities: Changing views on pair bonding and social organization in hominoids. *Yearbook of Physical Anthropology* 43: 33–60.

Fuentes A. 2002. Patterns and trends in primate pair bonds. *International Journal of Primatology* 23: 953–978.

Fuentes A. 2017. Human niche, human behaviour, human nature. *Interface Focus* 7(5): 20160136.

Furuichi T. 1987. Sexual swelling, receptivity, and grouping of wild pygmy chimpanzee females at Wamba, Zaire. *Primates* 28(3): 309–318.

Furuichi T. 1989. Social interactions and the life history of female *Pan paniscus* in Wamba, Zaire. *International Journal of Primatology* 10: 173–197.

Geissmann T. 2007. Status reassessment of gibbons: Results of the Asian Primate Red List Workshop 2006. *Gibbon Journal* 3: 5–15.

Geissmann T. 2008. Gibbon paintings in China, Japan, and Korea: Historical distribution, production rate and context. *Gibbon Journal* 4: 1–38.

Gittins S, Tilson R. 1984. Notes on the ecology and behaviour of the hoolock gibbon. In Preuschoft H, Chivers D, Brockelman WY, Creel N. (Eds.), *The Lesser Apes: Evolutionary and Behavioural Biology* (pp. 258–266). Edinburgh: Edinburgh University Press.

Goldsmith ML. 1996. *Ecological Influences on the Ranging and Grouping Behaviour of Western Lowland Gorillas at Bai Hokou, Central African Republic*. PhD Dissertation, State University of New York, Stony Brook.

Goodall J. 1986. *The Chimpanzees of Gombe: Patterns of Behaviour*. Cambridge, MA: Harvard University Press.

Gould SJ. 1994. The persistently flat earth. *Natural History*, 103(3): 12–19.

Gray JE (Ed.). 1870. *Catalogue of Monkeys, Lemurs, and Fruit-Eating Bats in the Collection of the British Museum*. London: British Museum Trustees.

Groves CP. 2001. *Primate Taxonomy*. Washington, DC: Smithsonian University Press.

Grueter CC, Ndamiyabo F, Plumptre AJ, Abavandimwe D, Mundry R, Fawcett KA, Robbins MM. 2013. Long-term temporal and spatial dynamics of food availability for endangered mountain gorillas in Volcanoes National Park, Rwanda. *American Journal of Primatology* 75(3): 267–280.

Haimoff EH, Yang JY, He SJ, Chen N. 1986. Census and survey of wild black crested gibbons (*Hylobates concolor concolor*). *Folia Primatologica* 46: 205–214.

Haimoff E, Yang XJ, He SJ, Chen N. 1987. Preliminary observations on black-crested gibbons (*Hylobates concolor concolor*) in Yunnan Province, People's Republic of China. *Primates* 28: 319–335.

Harcourt AH. 1979. Social relations among adult female mountain gorillas. *Animal Behaviour* 27: 251–264.

Harcourt AH, Stewart KJ. 2007. *Gorilla Society: Conflict, Compromise, and Cooperation between the Sexes.* Chicago, IL: University of Chicago Press.

Harrison T. 1982. *Small-Bodied Apes from the Miocene of East Africa.* PhD Thesis, University of London, London.

Harrison T. 1987. The phylogenetic relationships of the early catarrhine primates: A review of the current evidence. *Journal of Human Evolution* 16: 41–80.

Harrison T. 2002. Late Oligocene to middle Miocene catarrhines from Afro-Arabia. In Hartwig WC. (Ed.), *The Primate Fossil Record* (pp. 311–338). Cambridge: Cambridge University Press.

Harrison T. 2010. Dendropithecoidea, Proconsuloidea, and Hominoidea. In Werdelin L, Sanders WJ. (Eds.), *Cenozoic Mammals of Africa* (pp. 429–469). Berkeley: University of California Press.

Harrison T. 2013. Catarrhine origins. In Begun DR. (Ed.), *A Companion to Paleoanthropology* (pp. 376–396). New York, NY: Blackwell Publishing Ltd.

Harrison T. 2016. The fossil record and evolutionary history of hylobatids. In Reichard U, Hirai H, Barelli C. (Eds.), *Evolution of Gibbons and Siamang* (pp. 91–110). Developments in Primatology: Progress and Prospects. New York, NY: Springer.

Harrison T. 2017a. Miocene primates. *The International Encyclopedia of Primatology* 1: 1–5.

Harrison T. 2017b. Proconsul. *The International Encyclopedia of Primatology* 1: 1–2.

Hartwig W. 2011. Chapter 3: Primate evolution. In Campbell CJ, Fuentes A, MacKinnon KC, Bearder SK, Stumpf RM. (Eds.), *Primates in Perspective* (2nd ed., pp. 19–31). Oxford: Oxford University Press.

Hockings KJ, McLennan MR, Carvalho S, Ancrenaz M, Bobe R, Byrne RW, ... & Wilson ML. 2015. Apes in the Anthropocene: Flexibility and survival. *Trends in Ecology & Evolution* 30(4): 215–222.

Hohmann G, Fruth B. 2008. New records on prey capture and meat eating by bonobos at Lui Kotale, Salonga National Park, Democratic Republic of Congo. *Folia Primatologica* 79(2): 103–110.

Hohmann G, Gerloff U, Tautz D, Fruth B. 1999. Social bonds and genetic ties: Kinship, association and affiliation in a community of bonobos (*Pan paniscus*). *Behaviour* 136(9): 1219–1235.

Hon N, Behie AM, Rothman JM, Ryan KG. 2018. Nutritional composition of the diet of the northern yellow-cheeked crested gibbon (*Nomascus annamensis*) in north-eastern Cambodia. *Primates* 59(4): 339–346.

Hunt KD. 2016. Why are there apes? Evidence for the co-evolution of ape and monkey ecomorphology. *Journal of Anatomy* 228(4): 630–685.

Ishida H, Pickford M. 1997. A new late Miocene hominoid from Kenya: *Samburupithecus kiptalami* gen. et sp. nov. *CR Acad Sci D* 325: 823–829.

Jablonski NG. 2005. Primate homeland: Forests and the evolution of primates during the Tertiary and Quaternary in Asia. *Anthropological Science* 113: 117–122.

Jablonski NG, Chaplin G. 2009. The fossil record of gibbons. In Whittaker D, Lappan S. (Eds.), *The Gibbons: New Perspectives on Small Ape Socioecology and Population Biology*

(Chp. 7, pp. 111–130). Series Title: Developments in Primatology: Progress and Prospects. New York, NY: Springer Academic Press.

Kappeler M. 1984. Vocal bouts and territorial maintenance in the moloch gibbon. In Prueschoft H, Chivers D, Brockelman WY, Creel N. (Eds.), *The Lesser Apes: Evolution, Behaviour, and Biology* (pp. 376–389). Edinburgh: Edinburgh University Press.

Kay RF. 1977. Diets of early Miocene hominoids. *Nature* 268: 628–630.

Kelley J. 1997. Paleobiological and phylogenetic significance of life history in Miocene hominoids. In Begun DR, Ward CV, Rose MD. (Eds.), *Function, Phylogeny, and Fossils: Miocene Hominoid Origins and Adaptations* (pp. 173–208). New York, NY: Plenum Press.

Kendal J, Tehrani JJ, Odling-Smee J. 2011. Human niche construction in interdisciplinary focus. *Philosophical Transactions of the Royal Society B* 366: 785–792.

Kleiber M. 1975. Metabolic turnover rate: A physiological meaning of the metabolic rate per unit body weight. *Journal of Theoretical Biology* 53: 199–204.

Knott CD. 1999. *Reproductive, Physiological and Behavioral Responses of Orangutans in Borneo to Fluctuations in Food Availability*. Doctoral dissertation, Harvard University.

Knott CD, Beaudrot L, Snaith T, White S, Tschauner H, Planansky G. 2008. Female-female competition in Bornean orangutans. *International Journal of Primatology* 29: 975–997.

Knott CD, Kahlenberg S. 2011. Orangutans: Understanding forced copulations. In Campbell CJ, Fuentes A, MacKinnon KC, Bearder SK, Stumpf RM. (Eds.), *Primates in Perspective* (2nd ed., pp. 313–325). Oxford: Oxford University Press.

Koufos GD, de Bonis L. 2005. The Late Miocene hominoids *Ouranopithecus* and *Graecopithecus*. Implications about their relationships and taxonomy. *Annales de Paléontologie* 91(3): 227–240.

Lan D, Sheeran LK. 1995. The status of black gibbons (*Hylobates concolor jingdongensis*) at Xiaobahe, Wuliang Mountains, Yunnan Province, China. *Asian Primates* 5: 2–4.

Langergraber KE, Watts DP, Vigilant L, Mitani JC. 2017. Group augmentation, collective action, and territorial boundary patrols by male chimpanzees. *Proceedings of the National Academy of Sciences* 114(28): 7337–7342.

Lappan S. 2005. *Biparental Care and Male Reproductive Strategies in Siamangs (Symphalangus syndactylus) in Southern Sumatra*. PhD Thesis, New York University, New York.

Lappan S. 2007. Social relationships among males in multimale siamang groups. *International Journal of Primatology* 28: 369–387.

Lappan S, Whittaker DJ (Eds.). 2009. *The Gibbons: New Perspectives on Small Ape Socioecology and Population Biology*. New York, NY: Springer.

Larson SG. 1998. Parallel evolution in the hominoid trunk and forelimb. *Evolutionary Anthropology: Issues, News, and Reviews* 6(3): 87–99.

Lehmann J, Boesch C. 2008. Sexual differences in chimpanzee sociality. *International Journal of Primatology* 29(1): 65–81.

Leighton DL. 1987. Gibbons: Territoriality and monogamy. In Smuts B, Cheney DL, Seyfarth RM, Wrangham RW, Struhsaker TT. (Eds.), *Primate Societies* (pp. 135–145). Chicago, IL: University of Chicago Press.

Levins R, Lewontin R. 1985. *The Dialectical Biologist*. Cambridge, MA: Harvard University Press.

Liebal K. 2016. Communication and cognition of small apes. In Reichard U, Hirai H, Barelli C. (Eds.), *Evolution of Gibbons and Siamang* (pp. 91–110). Developments in Primatology: Progress and Prospects. New York, NY: Springer.

Liebal K, Waller BM, Burrows A, Slocombe K (Eds.). 2013. *Primate Communication: A Multimodal Approach*. Cambridge: Cambridge University Press.

Lodwick JL, Salmi R. 2019. Nutritional composition of the diet of the western gorilla (*Gorilla gorilla*): Interspecific variation in diet quality. *American Journal of Primatology* 81(9): e23044.

Lucchesi S, Cheng L, Janmaat K, Mundry R, Pisor A, Surbeck M. 2020. Beyond the group: How food, mates, and group size influence intergroup encounters in wild bonobos. *Behavioral Ecology* 31(2): 519–532.

Ma C, Liao J, Fan P. 2017. Food selection in relation to nutritional chemistry of Cao Vit gibbons in Jingxi, China. *Primates* 58(1): 63–74.

MacKinnon JR, MacKinnon KS. 1980. Niche differentiation in a primate community. In Chivers DJ. (Ed.), *Malayan Forest Primates: Ten Year Study in Tropical Rain Forest* (pp. 167–191). New York, NY: Plenum Press.

MacLeod CE, Zilles K, Schleicher A, Rilling JK, Gibson KR. 2003. Expansion of the neocerebellum in Hominoidea. *Journal of Human Evolution* 44: 401–429. Doi:10.1016/S0047-2484(03)00028-9.

Maggioncalda AN, Sapolsky RM, Czekala NM. 2002. Male orangutan subadulthood: A new twist on the relationship between chronic stress and developmental arrest. *American Journal of Physical Anthropology* 118: 25–32.

Malenky R. 1990. Ecological factors affecting food choice and social organisation in *Pan paniscus*. PhD Dissertation, SUNY at Stony Brook, New York.

Malone N, Fuentes A. 2009. The ecology and evolution of hylobatid communities: Proximate and ultimate considerations of inter- and intraspecific variation. In Whittaker D, Lappan S. (Eds.), *The Gibbons: New Perspectives on Small Ape Socioecology and Population Biology* (Chp. 12, pp. 241–264). Series Title: Developments in Primatology: Progress and Prospects. New York, NY: Springer Academic Press.

Malone N, Fuentes A, White FJ. 2012. Variation in the social systems of extant hominoids: Comparative insight into the social behavior of early hominins. *International Journal of Primatology* 33(6): 1251–1277.

Malone N, Selby M, Longo S. 2014. Political-ecological dimensions of silvery gibbon conservation efforts: An endangered ape in (and on) the verge. *International Journal of Sociology* 44(1): 34–53.

Marks J. 2012. Why be against Darwin? Creationism, racism, and the roots of anthropology. *Yearbook of Physical Anthropology* 55: 95–104. DOI:10.1002/ajpa.22163.

Martin RD. 1986. Primates: A definition. In Wood B, Martin L, Andrews P. (Eds.), *Major Topics in Primate and Human Evolution* (pp. 1–31). Cambridge: Cambridge University Press.

McGrew WC. 1992. *Chimpanzee Material Culture: Implications for Human Evolution*. Cambridge: Cambridge University Press.

McNulty KP. 2010. Apes and tricksters: The evolution and diversification of humans' closest relatives. *Evolution: Education and Outreach* 3(3): 322–332.

Mootnick AR, Haimoff EH, Nyunt-Lwin K. 1987. Conservation and captive management of hoolock gibbons in the Socialist Republic of the Union of Burma. *AAZPA 1987 Annual Proceedings* 398424: 398–423.

Morin PA, Moore JJ, Chakraborty R, Jin L, Goodall J, Woodruff DS. 1994. Kin selection, social structure, gene flow, and the evolution of chimpanzees. *Science* 265(5176): 1193–1201.

Moyá-Solá S. 2017. *Pierolapithecus*. *The International Encyclopedia of Primatology* 1: 1–3.

Moyá-Solá S, Köhler M. 1996. A *Dryopithecus* skeleton and the origins of great-ape locomotion. *Nature* 379: 156–159.

Mukherjee RJ, Chaudhuri S, Murmu A. 1991–1992. Hoolock gibbons (*Hylobates hoolock*) in Arunachal Pradesh, Northeast India: The Lohit District. *Primate Conservation* 12–13: 31–33.

Napier JR. 1970. Paleoecology and catarrhine evolution. In Napier JR, Napier PH. (Eds.), *Old World Monkeys: Evolution, Systematics, and Behavior* (pp. 53–95). London: Academic Press.

Nater A, Mattle-Greminger MP, Nurcahyo A, Nowak MG, De Manuel M, Desai T, … & Lameira AR. 2017. Morphometric, behavioral, and genomic evidence for a new orangutan species. *Current Biology* 27(22): 3487–3498.

Nishida T. 1990. *Chimpanzees of the Mahale Mountains.* Tokyo: University of Tokyo Press.

Nishida T, Hiraiwa-Hasegawa M. 1987. Chimpanzees and bonobos: Cooperative relationships among males. In Smuts B, Cheney DL, Seyfarth RM, Wrangham RW, Struhsaker TT. (Eds.), *Primate Societies* (pp. 165–177). Chicago, IL: University of Chicago Press.

Nishida T, Uehara S, Nyundo, R. 1979. Predatory behavior among wild chimpanzees of the Mahale Mountains. *Primates* 20(1): 1–20.

Oyama S, Griffiths PE, Gray RD. 2001. Introduction: What is developmental systems theory. In Oyama S, Griffiths PE, Gray RD. (Eds.), *Cycles of Contingency: Developmental Systems and Evolution* (pp. 1–12). Cambridge: MIT Press.

Palombit RA. 1992. *Pair Bonds and Monogamy in Wild Siamang (Hylobates syndactylus) and White-Handed Gibbon (Hylobates lar) in Northern Sumatra.* PhD Dissertation, University of California, Davis.

Palombit RA. 1994. Dynamic pair bonds in hylobatid: Implications regarding monogamous social systems. *Behaviour* 128(1–2): 65–101.

Palombit RA. 1996. Pair bonds in monogamous apes: A comparison of the siamang (Hylobates syndactylus) and the white-handed gibbon (Hylobates lar). *Behaviour* 133(5–6): 321–356.

Palombit RA. 1997. Inter- and intraspecific variation in the diets of sympatric siamang (*Hylobates syndactylus*) and lar gibbons (*Hylobates lar*). *Folia Primatologica* 68: 321–337.

Pilbeam D, Rose MD, Barry JC, Ibrahim Shah SM. 1990. New *Sivapithecus* humeri from Pakistan and the relationship of *Sivapithecus* and *Pongo. Nature* 348: 237–239.

Pilbrow V. 2006. Population systematics of chimpanzees using molar morphometrics. *Journal of Human Evolution* 51(6): 646–662.

Pina M, Almécija S, Alba DM, O'Neill MC, Moyá-Solá S. 2014. The middle Miocene ape *Pierolapithecus catalaunicus* exhibits extant great ape-like morphometric affinities on its patella: Inferences on knee function and evolution. *PloS ONE* 9(3). Doi.org/10.1371/journal.pone.0091944.

Potts R. 1998. Variability selection in human evolution. *Evolutionary Anthropology* 7: 81–96.

Potts R. 2004. Paleoenvironmental basis cognitive evolution in great apes. *American Journal of Primatology* 62: 209–228.

Preuschoft H, Demes B. 1984. Biomechanics of braciation. In Preuschoft H, Chivers D, Brockelman WY, Creel N. (Eds.), *The Lesser Apes: Evolutionary and Behavioural Biology* (pp. 96–118). Edinburgh: Edinburgh University Press.

Pruetz JD, Bertolani P, Ontl KB, Lindshield S, Shelley M, Wessling EG. 2015. New evidence on the tool-assisted hunting exhibited by chimpanzees (*Pan troglodytes verus*) in a savannah habitat at Fongoli, Sénégal. *Royal Society Open Science* 2(4): 140507.

Raaum RL, Sterner KN, Noviello CM, Stewart CB, Disotell TR. 2005. Catarrhine primate divergence dates estimated from complete mitochondrial genomes: Concordance with fossil and nuclear DNA evidence. *Journal of Human Evolution* 48: 237–257.

Raemaekers JJ. 1979. Ecology of sympatric gibbons. *Folia Primatologica* 31: 227–245.

Ramachandran VS. 2011. *The Tell-Tale Brain: A Neuroscientist's Quest for What Makes Us Human.* New York, NY: W.W. Norton & Company.

Reichard U. 1995. Extra-pair copulations in a monogamous gibbon (*Hylobates lar*). *Ethology* 100: 99–112.

Reichard UH. 2009. The social organisation and mating systems of Khao Yai white-handed gibbons: 1992–2006. In Lappan S, Whittaker DJ. (Eds.), *The Gibbons: New Perspectives on Small Ape Socioecology and Population Biology* (pp. 347–384). New York, NY: Springer.

Reichard UH, Barelli C, Hirai H, Nowak MG. 2016. The evolution of gibbons and sia-mang. In Reichard U, Hirai H, Barelli C. (Eds.), *Evolution of Gibbons and Siamang* (pp. 3–41). Developments in Primatology: Progress and Prospects. New York, NY: Springer.

Reichard UH, Ganpanakngan M, Barelli C. 2012. White-handed gibbons of Khao Yai: Social flexibility, complex reproductive strategies, and a slow life history. In *Long-Term Field Studies of Primates* (pp. 237–258). Berlin, Heidelberg: Springer.

Remis MJ, Dierenfeld ES, Mowry CB, Carroll RW. 2001. Nutritional aspects of western lowland gorilla (*Gorilla gorilla gorilla*) diet during seasons of fruit scarcity at Bai Hokou, Central African Republic. *International Journal of Primatology* 22(5): 807–836.

Robbins MM. 2003. Behavioural aspects of sexual selection in mountain gorillas. In Jones CB. (Ed.), *Sexual Selection and Reproductive Competition in Primates: New Perspectives and Directions* (pp. 477–501). New York, NY: American Society of Primatologists.

Robbins MM. 2011. Gorillas: Diversity in ecology and behaviour. In Campbell CJ, Fuentes A, MacKinnon, KC, Bearder SK, Stumpf RM. (Eds.), *Primates in Perspective* (2nd ed., pp. 326–339). Oxford: Oxford University Press.

Robbins MM, Bermejo M, Cipolletta C, Magliocca F, Parnell RJ, Stokes E. 2004. Social structure and life history patterns in western gorillas (*Gorilla gorilla gorilla*). *American Journal of Primatology* 64: 145–159.

Robbins MM, Robbins AM, Gerald-Steklis N, Steklis HD. 2005. Long-term domin-ance relationships in female mountain gorillas: Strength, stability, and determinants of rank. *Behaviour* 142: 779–809.

Rogers ME, Abernethy K, Bermejo M, Cipolletta C, Doran D, McFarland K, Nishihara T, Remis M, Tutin CEG. 2004. Western gorilla diet: A synthesis from six sites. *American Journal of Primatology* 64: 173–192.

Roos C. 2016. Phylogeny and classification of gibbons (Hylobatidae). In Reichard U, Hirai H, Barelli C. (Eds.), *Evolution of Gibbons and Siamang* (pp. 151–165). Developments in Primatology: Progress and Prospects. New York, NY: Springer.

Roth TS, Rianti P, Fredriksson GM, Wich SA, Nowak MG. 2020. Grouping behavior of Sumatran orangutans (*Pongo abelii*) and Tapanuli orangutans (*Pongo tapanuliensis*) living in forest with low fruit abundance. *American Journal of Primatology* 82(5): e23123.

Ruppell J. 2007. The gibbons of Phong Nha-Ke Bang National Park in Vietnam. *Gibbon Journal* 3: 50–54.

Russon AE, Wich SA, Ancrenaz M, Kanamori T, Knott CD, Kuze N, … & Sawang A. 2009. Geographic variation in orangutan diets. In Wich SA, Utami Atmoko SS, Setia TM, van Schaik CP. (Eds.), *Orangutans: Geographic Variation in Behavioural Ecology and Conservation* (pp. 135–156). Oxford: Oxford University Press.

Sheeran L. 1993. *A Preliminary Study of the Behavior and Socio-Ecology of Black Gibbons (Hylobates concolor) in Yunnan Province, People's Republic of China.* PhD Dissertation, University of Ohio, Columbus.

Sicotte P. 1993. Inter-group encounters and female transfer in mountain gorillas: Influence of group composition on male behaviour. *American Journal of Primatology* 30: 21–36.

Sicotte P. 1994. Effects of male competition on male-female relationships in bi-male groups of mountain gorillas. *Ethology* 97: 47–64.

Siddiqi NA. 1986. Gibbons (*Hylobates hoolock*) in the West Bhanugach Reserved Forest of Sylhet District, Bangladesh. *Tiger Paper* 8: 29–31.

Sommer V, Bauer J, Fowler A, Ortmann S. 2011. Patriarchal chimpanzees, matriarchal bonobos: Potential ecological causes of a *Pan* dichotomy. In *Primates of Gashaka* (pp. 469–501). New York, NY: Springer.

Srikosamatara S. 1984. Ecology of pileated gibbons in south-east Thailand. In Prueschoft H, Chivers DJ, Brockelman WY, Creel N. (Eds.), *The Lesser Apes: Evolution, Behaviour, and Biology* (pp. 242–257). Edinburgh: Edinburgh University Press.

Steffen W, Grinevald J, Crutzen P, McNeill J. 2011. The Anthropocene: Conceptual and historical perspectives. *Philosophical Transactions of the Royal Society of London A: Mathematical, Physical and Engineering Sciences* 369(1938): 842–867.

Steiper ME. 2006. Population history, biogeography, and taxonomy of orangutans (Genus: *Pongo*) based on a population genetic meta-analysis of multiple loci. *Journal of Human Evolution* 50: 509–522.

Sterck EHM, Watts DP, van Schaik CP. 1997. The evolution of female social relationships in nonhuman primates. *Behavioural Ecology and Sociobiology* 41(5): 291–309.

Stewart CB, Disotell TR. 1998. Primate evolution – in and out of Africa. *Current Biology* 8(16): R582–R588.

Stumpf RM. 2011. Chimpanzees and bonobos: Inter-and intra-species diversity. In Campbell CJ, Fuentes A, MacKinnon KC, Bearder SK, Stumpf RM. (Eds.), *Primates in Perspective* (2nd ed., pp. 340–356). Oxford: Oxford University Press.

Stumpf RM. 2017. Chimpanzee and bonobo (*Pan*). *The International Encyclopedia of Primatology* 1: 1–3.

Sugiyama Y. 2004. Demographic parameters and life history of chimpanzees at Bossou, Guinea. *American Journal of Physical Anthropology* 124(2): 154–165.

Sugiyama Y, Koman J. 1979. Tool-using and-making behavior in wild chimpanzees at Bossou, Guinea. *Primates* 20(4): 513–524.

Sussman RW. 1991. Primate origins and the evolution of angiosperms. *American Journal of Primatology* 23(4): 209–223.

Sussman RW, Cloninger RC. 2011. Origins of altruism and cooperation. *Developments in Primatology: Progress and Prospects* (Vol. 36, Part 2). New York, NY: Springer.

Sussman RW, Rasmussen TD, Raven PH. 2013. Rethinking primate origins again. *American Journal of Primatology* 75(2): 95–106.

Temerin LA, Cant JG. 1983. The evolutionary divergence of Old World monkeys and apes. *The American Naturalist* 122(3): 335–351.

Thinh VN, Mootnick AR, Thanh VN, Nadler T, Roos C. 2010. A new species of crested gibbon from the central Annamite mountain range. *Vietnamese Journal of Primatology* 4: 1–12.

Tilson RL. 1979. Behaviour of hoolock gibbon (*Hylobates hoolock*) during different seasons in Assam, India. *Journal of the Bombay Natural History Society* 79: 1–16.

Tutin CEG. 1996. Ranging and social structure of lowland gorillas in the Lopé Reserve, Gabon. In McGrew WC, Marchant LF, Nishida T. (Eds.), *Great Ape Societies* (pp. 58–70). Cambridge: Cambridge University Press.

Tyler DE. 1993. The evolutionary history of the gibbon. In Jablonski N. (Ed.), *Evolving Landscapes and Evolving Biotas of East Asia since the Mid-Tertiary* (pp. 228–240). Hong Kong: Center for Asian Studies, Hong Kong University.

Utami Atmoko SS, Wich SA, Sterck EHM, van Hooff JARAM. 1997. Food competition between wild orangutans in large fig trees. *International Journal of Primatology* 18: 909–927.

Van Gulik RH. 1967. *The Gibbon in China. An Essay in Chinese Animal Lore.* Leiden: E. J. Brill.

van Schaik CP. 1999. The socioecology of fission–fusion sociality in orangutans. *Primates* 40: 69–86.

van Schaik CP, Ancrenaz M, Borgen G, Galdikas B, Knott C, Singleton I, Suzuki A, Utami Atmoko SS, Merrill M. 2003. Orangutan cultures and the evolution of material culture. *Science* 299: 102–105.

Vigilant L, Bradley BJ. 2004. Genetic variation in gorillas. *American Journal of Primatology* 64: 161–172.

Wade A. 2020. *Shared Landscapes: The Human-Ape Interface within the Mone-Oku Forest, Cameroon.* PhD Thesis, University of Auckland.

Ward CV, Duren DL. 2002. Middle and late Miocene African hominoids. In Hartwig WC. (Ed.), *The Primate Fossil Record* (pp. 385–397). Cambridge: Cambridge University Press.

Warren KS, Verschoor EJ, Langenhuijzen S, Swan RA, Vigilant L, Heeney JL. 2001. Speciation and intrasubspecific variation of Bornean orang-utans, *Pongo Pygmaeus pygmaeus. Molecular Biology and Evolution*, 18(4): 472–480.

Watts DP. 1992. Social relationships of immigrant and resident female mountain gorillas, I. Male–female relationships. *American Journal of Primatology* 28: 159–181.

Watts DP. 2001. Female mountain gorillas: Social relationships. In Robbins MM, Sicotte P, Stewart KJ. (Eds.), *Mountain Gorillas, Three Decades of Research at Karisoke, Cambridge Studies in Biological and Evolutionary Anthropology* (Vol. 27, pp. 215–240). Cambridge: Cambridge University Press.

Watts DP. 2008. Tool use by chimpanzees at Ngogo, Kibale National Park, Uganda. *International Journal of Primatology* 29(1): 83–94.

Watts DP, Mitani JC. 2002. Hunting and meat sharing by chimpanzees at Ngogo, Kibale National Park, Uganda. In Boesch C, Hohmann G, Marchant LF (Eds.), *Behavioural Diversity in Chimpanzees and Bonobos* (pp. 244–258). Cambridge: Cambridge University Press.

White FJ. 1989. Social organization of pygmy chimpanzees. In Heltne PG, Marquardt LA. (Eds.), *Understanding Chimpanzees* (194–207). Cambridge, MA: Harvard University Press.

White FJ. 1994. Food sharing in wild pygmy chimpanzees, *Pan paniscus.* Social development, learning and behaviour. In Anderson JR, Herrenschmidt N, Roeder JJ, Thierry B. (Eds.), *Current Primatology Volume II: Social Development, Learning and Behaviour (Proceedings of the 14th International Primatological Society Congress)* (pp. 1–10). Strasbourg: Universite Louis Pasteur.

White FJ. 1996. *Pan paniscus* 1973 to 1996: Twenty-three years of field research. *Evolutionary Anthropology: Issues, News, and Reviews: Issues, News, and Reviews* 5(1): 11–17.

White FJ. 1998. Seasonality and socioecology: The importance of variation in fruit abundance to bonobo sociality. *International Journal of Primatology* 19(6): 1013–1027.

White FJ, Waller MT, Cobden AK, Malone NM. 2008. Lomako bonobo population dynamics, habitat productivity, and the question of tool use. *American Journal of Physical Anthropology* S46: 222.

White FJ, Wrangham RW. 1988. Feeding competition and patch size in the chimpanzee species *Pan paniscus* and *Pan troglodytes*. *Behaviour* 105(1–2): 148–164.

Whitmore TC. 1984. *Tropical Rainforests of the Far East*, 2nd ed. Oxford: Clarendon Press.

Whittaker DJ, Lappan S. 2009. The diversity of small apes and the importance of population-level studies. In Lappan S, Whittaker DJ. (Eds.), *The Gibbons: New Perspectives on Small Ape Socioecology and Population Biology* (pp. 3–10). New York, NY: Springer.

Wrangham RW. 1986. Ecology and social relationships in two species of chimpanzee. In Rubenstein DI, Wrangham RW. (Eds.), *Ecological Aspects of Social Evolution: Birds and Mammals* (pp. 352–378). Princeton, NJ: Princeton University Press.

Wu R, Pan Y. 1984. A late Miocene gibbon-like primate from Lufeng, Yunnan Province. *Acta Anthropologica Sinica* 3: 193–200.

Yamagiwa J, Kahekwa J, Basabose AK. 2003. Intra-specific variation in social organization of gorillas: Implications for their social evolution. *Primates* 44: 359–369.

Yi Y, Fichtel C, Ham S, Jang H, Choe JC. 2020. Fighting for what it's worth: Participation and outcome of inter-group encounters in a pair-living primate, the Javan gibbon (*Hylobates moloch*). *Behavioral Ecology and Sociobiology* 74(8): 1–15.

Zhenhe L, Yongzu Z, Haisheng J, Southwick C. 1989. Population structure of *Hylobates concolor* in Bawanglin Nature Reserve, Hainan, China. *American Journal of Primatology* 19: 247–254.

3 Emergence

Theorising ape sociality

> The study of the behaviour of nonhuman primates is also political. In hindsight, we can see the inscription of ideas about the family, gender relations, and even warfare upon animals who actually have no families, genders or wars but only metaphorical extensions of them.
>
> (Jonathan Marks 2009: 94)

The above quote represents an inherent tendency to interpret the results of materialist studies of primates through socio-centric lenses. In fact, as humans studying our closest relatives, there is no ability to completely escape our subjective tendencies. In response, behavioural researchers are rigorously trained to develop detailed ethograms and carefully crafted, operational definitions of behaviour in an attempt to eliminate any subjectivity whatsoever. On the other hand, as fellow hominoids, drawing upon our own perceptions can potentially bear insight into the social lives of apes. For example, when we observe conflict and post-conflict resolution behaviour, our intuitive read of the intensity and fall-out from inter-individual conflict may be highly accurate. Or, acknowledging how critical play behaviour and early socialisation are for successful development and maturation (even though the strength of such causal linkages would be nearly impossible to quantify in long-lived, wild primates).[1] Furthermore, in contemporary primatology, these tensions are increasingly heightened from the acknowledgement that (with near-ubiquity) humans and other primates co-reside in ecological and social landscapes (Fuentes 2010; McKinney 2015; Malone & Wedana 2019; Riley 2020). Some even posit that recent work in ethnoprimatology disproves "the idea that ecologies exist outside of the human realm" (Dore 2018: 937).

I agree that the ecological landscapes where we study primates are intertwined with, and to an extent determined by, social and political dynamics. This is certainly true at present, as well as in the historical and archaeological past, as both recorded and inferred interactions between humans and other primates are indicative of causal relationships on both ecological and evolutionary scales (Shipman et al. 1981; Tutin & Oslisly 1995; Harrison 1996; Ellwanger & Lambert 2018; Spehar et al. 2018). In fact, today we know that even cultural

DOI: 10.4324/9780367211349-3

beliefs and livelihood practices act as shaping factors of landscapes in ways that have vital implications for the resilience of both ecological and social systems. Contemporary landscapes do not exist independent of humans, nor do human societies exist independently of a biophysical context (Fuentes & Baynes-Rock 2017; York & Longo 2017). However, many aspects of nonhuman hominoid "ecologies" emerged prior to (i.e., independent of) human socio-political dynamics. These evolved baselines can act as both constraints and behavioural potentials.[2] They arise from, and exist within, a material reality. And herein lies the challenge for the development of predictive, socioecological models for primates in the anthropocene: we need to engage dialectically with the primates in our heads and the primates in the world.

And to do so we rely on theory. Routinely, our colleagues in anthropology and beyond draw heavily from the theoretical frameworks of political economy and political ecology (e.g., Rappaport 1968; Sahlins 1972; Escobar 1999).[3] In these frameworks, space is made for theorising the moments when the material and cultural aspects of the human niche are thrust into interaction. Early political ecologists felt that their approach offered the potential for a more socially just explication of narratives on environmental degradation by exposing the disparity in human–environment relationships and the historical contexts that led to them (Thompson & Warburton 1985; Blaikie & Brookfield 1987). From the analysis of historical events to the prediction of future outcomes, these theoretical models can help us to understand everything from colonialism to the management of endangered species. It strikes me that some anthropological primatologists might orient themselves towards a theoretical bearing I will refer to here as *critical political ethology*, or a realist-materialist approach to the animals in the world (following York & Longo 2017). In attempting to demonstrate the merit of such an approach for analysing behaviour, I return to the dialectical principles outlined in Chapter 1, and remain committed to finding a synthesis in the opposing states of "ape socioecology in the world" and the "ape socio-ecology in our heads". Throughout this chapter, which serves as both a review of the theoretical landscape in primatology and an attempt to open up new theoretical space, I will adhere to two guiding principles: (1) to take seriously the energetic and material exchanges between organism and environment and (2) to attend to all the mechanisms of evolutionary change and explore multiple channels of inheritance. I acknowledge that adherence to these principles predisposes my analysis to the discounting of hyper-reductionist and/or hyper-adaptationist explanations of socioecological phenomena.

Starting assumptions: Socioecological models and standard evolutionary theory

The morphological and behavioural diversity expressed by taxa within the order Primates (>500 species; 79 genera) is expansive. Furthermore, primates are essential ecological components of (primarily) tropical ecosystems, and display a stunning array of social system variants both within- and between-species

(Estrada et al. 2017; Strier 2017a). Research linking the underpinning ecological conditions to the various aspects of sociality (i.e., social organisation, social structure and mating patterns)[4] is a decades-old pursuit of primate behavioural ecologists, and the basis for the development of theoretical models (for reviews, see Kappeler & van Schaik 2002; Fuentes 2011; Parga & Overdorf 2011). In turn, the application of various socioecological models shapes the research agendas of primatologists towards the investigation of how social and behavioural phenomena are sensitive to an array of ecological variables (Clutton-Brock 1974; Wrangham 1979, 1980; van Schaik & van Hooff 1983; Sterck et al. 1997). In these conceptual models, factors such as the abundance, quality and distribution of food, as well as predation pressure, are thought to be determinative of group size and cohesion, feeding strategies, mating patterns, ranging behaviour and the nature of inter-individual and inter-group interactions. The socioecological paradigm not only posed interesting questions about how environmental factors, including the distribution of risks and resources, influence the way male and female primates behave, but also provided testable predictions derived from proposed hypotheses.

Underlying the aforementioned models' predictions is an explicit emphasis on the direct and indirect reproductive effects of an individual's actions (known as *inclusive fitness*). In short, these models compel practitioners on a (often explicitly stated and always implied) search for behavioural *adaptations* (i.e., traits favoured by natural selection and functioning to enhance the relative reproductive success of individuals). Additionally, many practitioners adhere to conceptualisation of genes as hyper-determinative, despite the fact that gene-trait linkages are exceedingly difficult to unravel (Goodman 2013), and especially so with respect to social behaviour. This elevation of Darwin and Wallace's mechanism (natural selection) to a position of primacy relative to the other mechanisms of evolutionary change (e.g., gene flow, genetic drift), in combination with the emphasis on gene-trait linkages (if only in principle), is known as the Modern Synthesis and forms the cornerstone of standard evolutionary theory (SET). Melinda Zeder (2017: 20160133) summarises this succinctly:

> Natural selection is recognised as the pre-eminent and ultimate causal force in evolution that sorts variation arising from random mutation, and passes on adaptive variations at a higher rate than less adaptive ones, resulting in an evolutionary process that proceeds at a gradual pace made up of small micro-evolutionary changes in the composition of individual genes and alleles within genes.

In the behavioural sciences generally, and in primatology specifically, the near-ubiquitous embrace of these principles by practitioners has resulted in the formation of a research paradigm.[5] This paradigm in primatology comprises three elements: (1) an emphasis on *natural and sexual selection*; (2) *reductionism* – a level of causation at the individual (or sub-organismal) level; and (3) differential proliferation vis-à-vis *competition* (Figure 3.1). In the context of a paradigm,

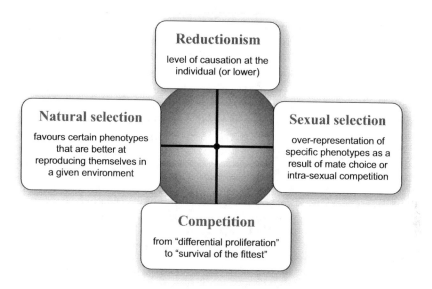

Figure 3.1 Axes of paradigmatic assumptions common in primatological research.

theorising can outpace data collection, and the reification of underlying assumptions can lead to explanations that may be theoretically sound but lacking in robust empirical support (e.g., dominance leads to resource holding potential and therefore selection favours behavioural propensities associated with aggression).[6] Two additional pervading assumptions include: (a) female mammals, investing more energy per offspring than males, are limited by, and therefore compete primarily for, nutritional resources; and (b) males, facing limited access to reproductive females, seek to maximise their own fitness (or minimise the variance) by balancing the costs and benefits of pursuing mates and engaging in intra-sexual competition (Williams 1966). For the past 50 years primatological practice has largely been conducted within such theoretical inertia.[7]

Theoretical and empirical limitations

It is clear that the intellectual histories of the biological and social sciences interact on a "two-way street of influence" (Longo & Malone 2006). Behavioural ecologists develop models and generate predictions (for humans and nonhumans alike) based on cost–benefit analyses of energy expenditure in relation to fitness returns (Maynard Smith 1982; van Schaik & Janson 2000; Krebs & Davies 2009). Meanwhile, some socioeconomic theories are based on rational individual choice that treats the maximisation of utility as the guiding principle of social relations. What is lost in this approach is the funda-mental conception of Darwinian evolution as a population-level process that examines changes in gene frequency in populations over time. That is to say, it is not an individual process per se. Population-level processes have much more complexity as they are interacting with a host of factors including environ-mental conditions, historical changes, density dependent effects and stochastic processes (such as genetic drift and some behavioural innovation) (e.g., Strum 2012). A dialectical approach does not ignore these influential factors. In fact, a dialectical analysis addresses such issues by taking into account the variety of interacting factors, and leads to a better understanding of social and biological processes. Giving ontological superiority to the individual over the social, what Marx referred to as "Robinsonades" (Marx 1973: 81), is an inherently non-dialectical approach (Longo & Malone 2006). For primatologists, a parallel can be found in expectations that an individual study subject will behave optimally, detached from the particular social and ecological histories of the society.

In primatology, while it is acknowledged that "the social organisation and demographic conditions created by individual behaviours also impose constraints on the behavioural options of these same individuals, leading to complex feedback loops" (Kappeler & van Schaik 2002: 708), individual research projects are rarely designed to identify and account for these dynamics. In long-lived species such as primates, even proxies of reproductive success are challenging to obtain, let alone direct measures of survivorship and life-time fitness (Ellwanger & Lambert 2018). Even variation within tried and true proxies (e.g., copulation rates; indices of social network strength) may have only limited correlational strength with the range of behavioural variance present within a population. Should this give us pause? Are we in fact recording optimal behavioural strategies?

Overall, the results of species-specific investigations of primate behavioural ecology are equivocal: some have found concordance among predictions from the synthetic socioecological model and empirically documented primate behaviour while others have failed to do so (succinctly reviewed by Riley 2020). Additionally, broad quantitative analyses document correlations between behavioural and eco-logical variables by comparing data from closely related taxa living in different environments, distantly related taxa living in similar environments and more recently among populations within the same taxa (Clutton-Brock 1974; Clutton-Brock & Harvey 1977; Strier 2003; Kamilar & Ledogar 2011). Pivoting away from strict ecological determinism, phylogenetic signals are sought via comparisons of related taxa (Di Fiore & Rendall 1994). However, the greatest challenge to a robust, predictive model in primatology stems from an acknowledgement of wide-spread intraspecific behavioural variation. Strier (2017b: 4–5) states:

> most contemporary primatologists now recognise that most primate species exhibit high levels of intraspecific variation in many facets of their behav-iour that do not always align with genetic or ecological differences and that may or may not have adaptive value.

Indeed, as we refine our research methods, and the overall number of studies increases, we are faced with both a broad range of social-system-wide variants between taxa, and also increasingly prevalent within-taxa "variations on a theme" (Strier 1994; Thierry 2008). Bernard Thierry (2008) goes as far as to suggest that primatologists should consider discarding the current socioecological approach and its expectation that a single comprehensive model will be able to accurately describe and predict primate social systems. My personal stance is that a significant revision is required rather than outright rejection (*sensu* Koenig & Borries 2009; Clutton-Brock & Janson 2012). I have also suggested that the ubiquitous nature of variation within social systems leads to a paradox whereby our interpretations are simultaneously more informed yet less absolute (Malone et al. 2012). The traditional socioecological model does not, inherently, incorporate substantial variation and social flexibility as core features of social systems.

In the next section, I offer a mode to ameliorate this paradox, with respect to the extant hominoids, by considering the evolutionary processes underlying behavioural variation. Importantly, Strier (2017b) differentiates between the behavioural variation of a species (where components of the social system vary with local ecological and/or demographic conditions) and behavioural flexibility of individuals in response to ecological or social stimuli. The latter, behavioural flexibility, occurs on temporal scales within the course of an individual's lifetime, and is theoretically reversible in absence of the transient pressure (Strier 2017b). Moreover, Piersma and Drent (2003: 28) suggest that "when environments change over shorter timescales than a lifetime, individuals that can show continuous, but reversible, transformations in behaviour, physiology or morphology, might incur a selective advantage". If we are to embrace these perspectives, we will have to consider not only evolutionary constraints (e.g., Foley & Lee 1989; Thierry 2008), but also alternative evolutionary pathways in addition to standard trait selection. An important dynamic that I wish to consider is the interplay of two classes of variation: (1) the evolved, taxon-specific behavioural variance; and (2) flexibility in individual response. To accomplish this, we will need to make space within an evolutionary model for *emergence* (dynamics among various levels of organisation) and *constrained totipotentiality* (the channelling of intraspecific variation); and this requires us, in the first instance, to follow recent calls to move beyond SET. If successful, we should be able to model more dynamic processes and better explain the observed variation and flexibility present in extant ape societies.

Principles of the extended evolutionary synthesis

Individuals are a critical unit of analysis in evolutionary biology, and especially within SET. Individuals vary, develop, migrate, reproduce and form social groups. The mechanisms of evolutionary change are widely considered to act primarily (though not exclusively) upon individuals.[8] With enhanced sociality considered to be a core, primate-wide evolutionary trend, the mediation of complex sociality by individuals is evolutionarily consequential, and its analysis represents the core business of many primatologists. As mentioned previously, the logic of competitive regimes and cost/benefit calculations are part of a package of basal assumptions. In this light, even the elucidation of *cooperative behaviour*, accepted as a vital and prevalent component of primate sociality, presents theoretical and empirical challenges (Sussman et al. 2005; Nowak 2006; Langergraber et al. 2007). Here again, we find ourselves pushing against the limits of reductionism. When scientific endeavours atomise and reduce natural phenomena in order to fit them into a conception of the world where parts have ontological superiority to the whole, a distorted view develops that can be described as alienated from the larger ecological process. As Levins and Lewontin (1985) explain, this alienated view is both ideological and real. The claim that the social order is the natural result of competing interest groups is an ideological formation intended to make the structure seem inevitable, but it also reflects the reality that has been constructed (Longo & Malone 2006). Moreover, a reductionist approach, while sometimes useful for understanding the workings of parts of a system, limits the scope of the analysis by developing dualisms or dichotomies, as opposed to understanding whole processes, missing the interaction and interdependence of parts that are crucial to the system (Gould & Lewontin 1979). The dichotomies that develop (e.g., individual or group; competition or cooperation) can be counterproductive and are often more reflective of social relations (in the production of scientific knowledge) rather than biological realities.[9]

In the interest of moving forward towards a reconciled, theoretical position, I briefly reprise the foundational principles of the dialectical primatologist, as introduced in Chapter 1. First, the rejection of an exclusively reductive view of biology through the explicit consideration of historical, emergent and contingent processes. Second, incorporating *totality* in behavioural ecology, or taking seriously the relational nature of categorical terms (e.g., "individual" and "group"; "organism" and "environment"). In addition to the contributions of Gould, Lewontin and Levins previously emphasised, a great deal of recent work by evolutionary theorists has resulted in a growing call to significantly revise the Modern Synthesis in order to account for new principles and processes (e.g., Laland et al. 2015; Zeder 2017).[10] This perspective involves the incorporation of dynamic evolutionary processes *in addition* to the action of natural and sexual selection. Such processes facilitate the potential for evolutionarily relevant, but non-genetically determined, insertions of pertinent behavioural patterns (West-Eberhart 2003; Jablonka & Lamb 2014). Previously, along with my colleagues Agustín Fuentes and Frances White, I argued that these insertions do not necessarily stem from "optimal adaptive trajectories as measured by models of reproductive fitness, but rather in broader behavioural potentials and social systems that diminish the pressure for, and impact of, individual strategies of cost minimisation and benefit maximisation" (Malone et al. 2012: 1253). Selection and other evolutionary factors mould aspects of the genetic, morphological and behavioural variation in primate populations, but the day-to-day behavioural interactions among primates also actively shape the social environment. The variation we see in the structures of ape societies lends itself to a diverse array of social bonds and networks (see Chapter 2). Rather than assuming a drive towards optimal solutions, I suggest that the interface between organisms and their environments should also be viewed as historical, population-level processes. Let me reiterate, I am not calling for a wholesale rejection of socioecological models and the orthodox evolutionary principles that underlie them, but rather for a significant revision (Table 3.1). However, given that our attention here is somewhat narrowly focused upon the Hominoidea, I am making the specific argument that SET fails to account for and/or predict the complexity of sociality within the superfamily.

In pivoting towards an extended evolutionary synthesis (EES), a key point of emphasis is in understanding the relationship between variation and evolutionary mechanisms. In SET, natural selection shapes variation that has arisen with no preferred orientation towards adaptive directions (i.e., "random", although not in the sense of mathematical probabilities). Rather, selection works upon "un-oriented" variation and changes the population via the enhanced fitness of individuals with such advantageous variants (Gould 1992). Contrastingly, niche construction theory (NCT), a core component of the EES, focuses on the coevolution of organisms and environments. Niche construction is "the process whereby organisms, through their metabolism, their activities, and their choices, modify their own and/or each other's niches"

Table 3.1 Comparison of parameters and predictions between SET and EES

	Standard Evolutionary Theory	*Extended Evolutionary Synthesis*
Causality	Uni-directional; organisms shaped primarily by natural selection	Reciprocal; organisms shape, and are shaped by, developmental environments
Directionality	Variation arises randomly, or is un-oriented	Variation arises through a combination of genetic and constructive processes
Targets of Selection	Alleles and genes	Alleles and genes; organisms and/or groups of organisms
Inheritance	Genetic inheritance; genetically encoded traits	Both genetic and non-genetic channels; ecological inheritance
Social Systems	Determined by the summation of individual competitive strategies	Social solutions to ecological problems; cohesion/ cooperation advantageous; extra-group context important
Pace of Change	Gradual and incremental	Both periods of stasis and rapid change

Source: See Laland et al. (2015); Futuyma (2017); Zeder (2017); Baedke et al. 2020.

(Odling-Smee et al. 2003: 419). Furthermore, organisms potentially pass along these altered environments to subsequent generations. The critical component of niche construction theory is that while genetic variation is subject to natural selection via differential survival and reproductive success, the very selective environments themselves are also engaged and structured by modifications made by niche constructing organisms (Kendal et al. 2011). Organisms' bias selection pressures to suit themselves by way of reciprocal, causal bouts of natural selection and niche construction resulting in a form of *constructed compatibility*, or adaptive complementarity. This perspective models both organismal and environmental change as a mutually mutable function of the interaction between organism × environment; a dynamic interface affecting the shape and behaviour of populations and the patterns and characteristics of selective pressures in ecosystems.

Odling-Smee et al. (2003) identify four major consequences of niche construction. Niche construction: (a) impacts/alters energy flows in ecosystems through ecosystem engineering (the creation and/or modulation of biotic or abiotic habitat parameters); (b) demonstrates that organisms modify their, and other, organisms' selective environments; (c) creates an ecological inheritance, including modified selection pressures, for subsequent populations; and (d) is a process, in addition to natural selection, that contributes to changes over time in the dynamic relationship between organisms and environments (niches). Recognising an ecological inheritance, in addition to genetic inheritance, serves to de-couple biological evolution from strict, inter-generational replication. Flack et al. (2006) make a compelling case that social networks constitute essential resources in gregarious primate societies.[11] The non-kin alliances that are important for chimpanzees, the homo- and hetero-sexual bonding in gorillas and the community-level associations among gibbon groups are all indicators of the importance of social networks, and demonstrate the potential of plasticity in the behavioural systems of the hominoids. Individuals in hominoid groups negotiate these social networks by modifying their boundaries and internal landscapes in the context of changing social relationships and demographics. The flexible social and ecological characteristics of the hominoids (within certain limitations) can act as a niche constructing process that both maintains flexibility and modifies selection pressures that then feedback on the system modifying the patterns that the variation takes (Malone et al. 2012).

As previously articulated in Malone et al. (2012), the basic premise of the model is that the early-mid Miocene hominoids exhibited a limited range of social plasticity relative to modern hominoids, but, like most anthropoid primates, had an emphasis on social bonding between select individuals. This led to the creation and maintenance of social networks within groups that emerged as core facets of their social system. Over the course of the Miocene (as discussed in Chapter 2), variability in social and ecological selection pressures favoured those individuals who were able to exhibit increased plasticity in behavioural response as a means to ameliorate external pressures. This selection for plasticity in behavioural response combined with slow rates of maturation and increased pressure for maintenance of complex social relationships would have enabled expansion in neurological elements (e.g., von Economo and mirror neurons, e.g., Allman et al. 2010; Ramanchandran 2011). These mutually constructed ecological and social pressures favoured a wider range of potential behavioural response without requiring the selection of specific behavioural patterns. With increased social and ecological complexity, social groups (and the individuals within them) experienced more evolutionarily relevant non-genetic inheritance (ecological and social). This, combined with expanded abilities to respond to selection pressures via relatively plastic behavioural responses or innovations, increased the role of a feedback system between niche construction and selection pressures favouring continued expansion in social plasticity via behavioural and neural expansion (MacKinnon & Fuentes 2011; Malone et al. 2012; Sueur et al. 2019) (Figure 3.2).

Critical political ethology: Elucidating ape sociality

In the previous chapter, I reviewed the social systems of the extant members of the superfamily Hominoidea. To some degree, individuals and groups from all ape taxa possess the ability and tendency to fission and fusion temporally and spatially while maintaining social cohesion across a range of ecological circumstances. As such, ape social relationships extend beyond kin and beyond the immediate social group. Niche construction is occurring in the hominoids, so we can envision the flexibility present at multiple levels as constituting a major component of the hominoid niche. Together with the fact that hominoids possess "a basic cognitive capacity for solving nonsocial problems with social solutions" (Reichard 2009: 371). This suggests the functioning of an adaptive social niche that is both inherently and structurally flexible.

An additional consideration that requires mention here, but will be more completely unpacked in subsequent chapters, is the small matter of the approximately eight billion human beings that presently (or will shortly) occupy the planet.[12] It is now increasingly understood that in addition to the distribution of risks, resources and the competitive regimes of conspecifics, primate social systems are subject to the influences of past and present anthropogenic alterations (Campbell-Smith et al. 2011; Hockings et al. 2015; Lappan et al. 2020). Anthropogenic patterns of ecological disturbance, including habitat degradation and hunting pressure, drive ecological, and subsequently social systems, into states of disequilibrium, sometimes leading to higher densities and larger group sizes, and sometimes leading to declines in density and group sizes (Struhsaker 1997, 1999). Actively perceiving that humans and other primates co-reside in ecological and social landscapes is now recognised as both a necessary and efficacious approach in primatology. Studies across a variety of taxa have successfully documented the development of primate behavioural and ecological strategies in response to habitat alterations by humans (e.g., Strum 2010; Hockings & McLennan 2012; Behie et al. 2019). A consequence of disturbance is the indirect manipulation of ecological and demographic variables

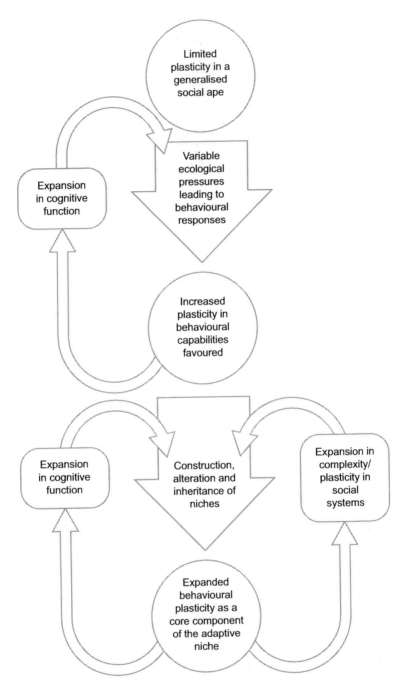

Figure 3.2 Simple trajectory for expanding social plasticity. Specific aspects of cognition, especially those related to behavioural plasticity, evolved largely in response to the increasingly challenging demands of a social life that includes the construction, alteration and inheritance of social and ecological niches (Malone et al. 2012).

that may favour phenotypic flexibility. The interplay of human and nonhuman primate ecologies creates a complex dynamic that can be characterised as a form of co-participatory niche construction (Riley & Fuentes 2010; Ellwanger & Lambert 2018).

It is also my assertion that either preconceptions about ape sociality, or an *a priori* predilection for adaptationism, or both, have in some cases biased the interpretation of ape behaviour. If my analyses of ape evolution (Chapter 2) and extant ape socioecology (Chapters 2 and 3) hold water, then I contend that much of the variation evident in extant hominoid social behaviour reflects epiphenomenal adaptive response (plasticity in expressed social behaviour) to local environmental and social conditions (i.e., *not* optimally fixed behavioural patterns produced by selection) (see also Malone et al. 2012). Indeed, such responses are expected within an "inclusive inheritance" framework whereby multiple mechanisms contribute to heredity including the transference of genetic, behavioural and ecological resources from parent to offspring (Jablonka & Lamb 2014; Laland et al. 2015). Keeping the workings of these intertwined dynamics in mind, I will now demonstrate how this approach can re-frame the analysis of three exemplary, social phenomena in a subset of the living apes: (1) hylobatid community dynamics; (2) differences in the social structure of bonobos and chimpanzees; and (3) orangutan social flexibility and resilience. In the three case studies that follow, I combine principles of the extended evolutionary synthesis with the critical lens of a dialectical primatologist.

Hylobatid community dynamics

The past 20+ years of discourse on hylobatid sociality has witnessed a marked shift.[13] The characterisation of hylobatids as obligate monogamous primates has largely been abandoned in favour of a view that identifies a considerable degree of social flexibility in the various components comprising the social system (Fuentes 2000; Malone & Fuentes 2009; Reichard et al. 2016). The nature of the debate can partially be attributed to the history of priorities within the discipline that can be characterised as somewhat back-to-front: that is deductive, hypothesis-driven approaches were advanced prematurely, and prior to the realisation of sufficiently robust, long-term natural history data (Sussman 2011). Until recently, the analysis of gibbon sociality has focused on testing hypotheses derived from sexual selection theory. It is possible that, couched in a paradigm (as described earlier in the chapter) that seeks to develop explicitly adaptationist models to explain the characteristic features of hylobatid social organisation, our models became "uprooted" from the ecology and behaviour of our study subjects (Malone & Fuentes 2009). For example, under such a rubric, practitioners may be more inclined to model fitness maximisation schema responsive to intersexual parental investment than to consider the group size and ranging limitations of ecological parameters, as well as the overall patterning of inter-group behaviour (Bartlett 2011; but see Robbins et al. 1991). When researchers initiate a study using a basal set of assumptions about the systems being studied, interpretations of collected data have the potential to be alienated from the actual ecological and behavioural processes involved (Longo & Malone 2006). In the case of gibbons, their classification as "monogamous" primates resulted in a specific suite of accompanying assumptions about their ecological and behavioural adaptations that potentially inhibited a broader investigation into behavioural variation across the full range of hylobatid taxa.

What we now know about gibbons (with respect to social systems and behavioural flexibility) has changed substantially in the past two decades. Specifically, we now recognise that individual gibbons are connected reproductively and behaviourally to a broader community beyond the immediate social group (Lappan & Whittaker 2009; Reichard et al. 2012). It is also increasingly understood that ecological pressures, in addition to intra-sexual competition, drive core aspects of gibbon social systems (e.g., relatively small group sizes and close socio-spatial proximity among gibbon pairs) (Bartlett 2009; Malone & Fuentes 2009; Chapter 2). In my own study site on the south coast of West Java, the Cagar Alam Leuweung Sancang (CALS, or the Sancang Forest Nature Reserve), an isolated population of silvery gibbons (*Hylobates moloch*) has been closely monitored for more than a decade (2005–2016) (Malone 2007; Malone & Wedana 2019). Within CALS there are fragments of unequal sizes and gibbon group densities. Specifically, a fragment of approximately 200 hectares and a second fragment of 400 hectares are inhabited by six and two gibbon groups, respectively (Malone & Oktavinalis 2006; Malone 2007; Reisland 2013). This inverse relationship between fragment size and the number of gibbon groups (despite identical habitat parameters) can only be understood historically in the context of forest loss, territorial size and group compositions, as well as inter-group relationships. In a compelling and complementary study, Reisland and Lambert (2016) observed two groups (B and C) in CALS over a period of 10 months and found Group B to be more sensitive to the presence of humans than Group C, occupying their range in non-random patterns to avoid humans which they may perceive as risky. In contrast, Group C's behaviour did not support the hypothesis that humans were a risk to be avoided, and this group was just as likely to be observed within areas of intensifying human presence as in areas without humans (Reisland & Lambert 2016). Interestingly, it is Group C that has demonstrated a more stable succession of individuals due to births, maturation and dispersal (Malone & Wedana 2019). Varying experiences with humans are possible for these groups because in some parts of the reserve humans represent both a real and a perceived risk to gibbons, whereas elsewhere, the presence of spiritual sites and practices seems to limit detrimental human activities with potential advantages for gibbon population viability (more on this in Chapter 4).

At CALS, we see an active and emergent relationship between individuals, groups, populations and their environments. This perspective is compatible with an increasing acknowledgement in evolutionary biology that a mutually interactive relationship exists between organisms and their environment through niche construction (Laland et al. 2001; Odling-Smee et al. 2003; Kendal et al.

2011). Niche construction conceptualises evolution of *both* populations of organisms and their environments through a dynamic, mutually mutable process that shapes the behaviour of organisms and the patterning of selective pressures in ecosystems (Odling-Smee 2007). The ability for organisms to not only impact their environment, but also, in part, shape the selective forces that they face, may result in more plastic or variable behavioural profiles (West-Eberhardt 2003; Malone et al. 2012). In the case of gibbons, the construction of territories and territorial relationships among contiguous groups creates a social and ecological niche relationship via genetic and ecological inheritance systems. At CALS, these processes are further influenced by the activities of humans, including (though not limited to): (a) cultural beliefs about the sacredness of the forest; (b) the consistent human presence that variably impacts the range use of gibbon groups (Malone 2007; Reisland & Lambert 2016); and (c) acute, intermittent human activities that directly impact the demography of the gibbon population. A model for population-level change over generational time that includes all of these factors will result in a comprehensive understanding of feedback pathways within the system (Figure 3.3). If this view of the variability is more widespread within the Hylobatidae, it suggests that we need to envision a broader evolutionary scenario. Expanding the focus beyond mating monogamy, pair-bondedness and territoriality, compels us to incorporate theoretical toolkits that go beyond basic selection models. As such, it is clear that the combination of a solitary evolutionary mechanism (i.e., selection, writ large) acting in response to a singular pressure (e.g., limited resources; infanticide) fails to predict the social variability in the Hylobatidae. In a final note, these debates are more than theoretical as conservation practitioners (e.g., the designers of rehabilitation and reintroduction protocols) may be hesitant to

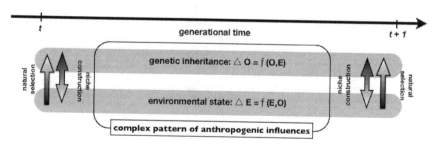

Adapted from Odling-Smee 2007

Figure 3.3 A model for socioecological change through time that combines the effects of natural selection, niche construction and the locally specific, salient patterns of anthropogenic influences. Niche construction theory conceptualises evolution of *both* populations of organisms (O) *and* their environments (E) through reciprocal, causal bouts of natural selection (NS) and niche construction (NC) over generational time (Malone & Wedana 2019, adapted from Odling-Smee 2007).

embrace progressive views of ape sociality without a concomitant increase in field-based, empirical validation.

Differences in the social structure of bonobos and chimpanzees

Set within the fission-fusion communities of chimpanzees and bonobos, we find behavioural traditions, boundary patrols, tool creation/use, and dynamic inter- and intra-group power differentials. Arguably, in these two *Pan* species, we find the clearest evidence for the presence of social and ecological niche construction, outside of humans. Differences between chimpanzees and bonobos are indeed evident, particularly in the strength of social alliances among male chimpanzees and, conversely, among female bonobos. However, as both intra-specific behavioural variation and ecological parameters vary dramatically throughout their respective ranges (more so for chimpanzees), a tightly linked co-variation between the two has yet to be conclusively determined (Sommer et al. 2011). Here, I will address what has come to symbolise as the most salient behavioural difference between chimpanzees and bonobos: patterns of aggression and lethal violence. Interest in this particular interspecific difference stems from the powerful influence, in both research circles and public imaginations, of perceived "warlike" or "peaceful" tendencies in our closest living relatives (Fry 2013). In 2014, Michael Wilson and colleagues published a study in the journal *Nature* declaring support for the hypothesis that lethal violence by chimpanzees is the result of adaptive strategies. Science writer Jonathan Webb proclaimed "Murder 'comes naturally' to chimpanzees" (*BBC News*, 18 September 2014). Various other headlines from around the world echoed this sentiment.[14] By compiling data across species, sites and observation years, Wilson et al. (2014) documented a difference between species in instances of lethal aggression (58 observed, plus a further 94 suspected or inferred cases in chimpanzees, versus one suspected killing in bonobos). Furthermore, the authors dismissed the possibility that human interventions (such as provisioning animals for research purposes or the alteration of habitats) were responsible for the rates of lethal violence between (and within) chimpanzee communities.[15]

One of the more compelling explanations of the varying "natures" within the genus *Pan* was advanced by Brian Hare, Victoria Wobber and Richard Wrangham in *Animal Behaviour*. Hare et al.'s (2012) model of "bonobo self-domestication" sets out to explain the numerous differences (behavioural, morphological and physiological) between bonobos and chimpanzees arising via a process analogous to the domestication of mammalian species by humans. That is, bonobo characteristics relative to chimpanzees are akin to the differences between domestics such as dogs, experimental foxes and guinea pigs and their respective wild progenitors. The down-regulation of aggression is further associated with developmental trajectories that produce a suite of traits, components of the so-called self-domestication syndrome, that emerge as epiphenomenal by-products (i.e., not adaptations, per se). An essential component of the self-domestication hypothesis is the role of consistent and strong selection pressures favouring reduced aggression. To the point, Hare et al. (2012: 579) state: "selection would have been strongest against lethal male aggression (e.g., infanticide) and might have operated through female choice given the cost that female primates pay as a result of male aggression" (citing Engh et al. 2006). The self-domestication hypothesis (SDH) rests upon two key assumptions: (1) that the frequency and intensity of intra- and inter-group aggression represent clear differences in the behavioural totipotentiality of bonobos and chimpanzees; and (2) that the last common ancestor (LCA) of chimpanzees and bonobos was "chimpanzee-like" and, since their divergence, bonobos have undergone intense, directional selection for reduced aggression.

Let's start with the first assumption. Are bonobos selected to be less aggressive than chimpanzees? In short, no. Bonobos have the potential to be highly aggressive, both in captivity and in the wild. Inter-individual conflict within captive colonies of bonobos is of sufficient intensity for participants to incur significant injuries (FJ White & K Boose, personal communication). In the wild, Hohmann and Fruth (2011) describe in detail a remarkable case of intra-group lethal violence against an adult male ("Volker") within the Lomako Forest, DRC study site. And in August 2011, three Congolese trackers were viciously attacked by rehabilitated orphan bonobos at a release site near Basankusu. Two of the trackers required emergency medical care and transport to France for facial reconstructive surgery (Andre 2011). Yes, the overall rates and intensity of aggressive interactions (both within- and between-communities) are reduced, though not absent, in bonobos as compared to chimpanzees (Sommer et al. 2011; Stumpf 2017). All age/sex classes of bonobos can be on either side of aggressive interactions, and the use of aggression can be an effective tool to stabilise the social network (Flack et al. 2006; Boose & White 2017). Bonobos are not selected to be less aggressive, but aggression may not serve them well in most circumstances. Given the complex intersexual rank relationships in bonobos, the social context is critical. Coalitions of females suppress and/or punish intra-group aggression (Kano 1992; Furuichi 2011). The greater female affiliation in bonobos may be possible through reduced feeding competition and the consumption of terrestrial herbaceous vegetation (Wrangham 1986), or the presence of larger feeding patches associated with the occupation of more productive habitats (Wrangham & White 1988; White 1996; Tokuyama et al. 2019).

The second assumption of the SDH concerns the nature of the LCA of chimpanzees and bonobos, and what happened to the lineages subsequent to divergence some 1–2 mya. From what we know about hominoid evolution, what evidence is there that the LCA of chimpanzees and bonobos was in fact "chimpanzee-like"? It is an equally plausible alternative that the LCA was "bonobo-like". Hare et al. (2012) acknowledge this alternative scenario, but dismiss it (tenuously) by inferring that cranial development in chimpanzees represents the ancestral African ape state (relative to the derived state of bonobo crania), and that behavioural reflections of this morphology can be inferred

(Wrangham & Pilbeam 2001). More importantly, either of those scenarios require strong and consistent directional selection for a fairly specific behavioural complex. I would argue that my overview of hominoid evolution (Chapter 2) supports a more conservative and parsimonious scenario, whereby the LCA of extant *Pan* would have been a more generalised African ape comprised of behavioural trait mosaicism. From this common ancestor, the specific ecological conditions, population demographics and history of social exchanges, generation on generation, would produce behavioural divergence between the chimpanzee and bonobo lineages. There is no "one way" to be a chimpanzee or a bonobo. Rather, general taxon-specific biologies can be viewed as platforms for a wide range of behavioural possibilities.

Does niche construction and social/ecological feedback better explain the social systems of chimpanzees and bonobos than a singular, directionally consistent and strong selective pressure? Social niche construction, and ecological inheritance in general, can occur via behavioural action, and I suggest that we are observing these mechanisms at play in the Hominoidea (generally) and in *Pan* (specifically) vis-à-vis their socially and ecologically flexible responses to selective pressures. Additionally, the potential for innovation, spread and inheritance of behavioural patterns and social traditions is much higher in the hominoids than in other anthropoid primates. While in most non-hominoid primates, variability and flexibility take the shape of specific patterns of demographic flux and inter-individual relationships, such as basic alliances and coalitions, we should consider behavioural flexibility and plasticity as "means-to-an-end" in hominoid socioecological landscapes. I see this emergence of behavioural variation as broadly similar to the expression of "reaction norms" (McNamara & Houston 2009; Dingemanse et al. 2010), where socioecological context dependency selects for behavioural ranges rather than singular types. A difference lies within the evolutionary mechanism(s) involved, and the opening-up of multiple channels of inheritance. In my view, a more robust mechanism is found in the multiple modes of inheritance made explicit by niche construction and the EES (Jablonka & Lamb 2008, 2014), as compared to a more limited genetic response to socioecological determinants (Jaeggi et al. 2016).

Orangutan social flexibility

In Chapter 2, I presented an overview of the genus *Pongo*. To summarise: though often characterised as leading a semi-solitary existence, orangutan females (and immature offspring) form social groupings that consist of travel bands, feeding aggregations and consortships with individual males (Utami et al. 1997; Knott et al. 2008). Male ranges are large and overlap with both male and female ranges. Perhaps this is best described as dispersed sociality, or an individual-based fission–fusion system (van Schaik 1999). Additionally, in areas of higher-than-usual/consistent productivity (e.g., Suaq Balimbing, Ketambe and Sikundur,

Sumatra), as well as in captive and provisioned reintroduction sites, orangutans appear to be particularly gregarious (Roth et al. 2020). Especially relevant to the discussion here, behavioural transmission and diffusion of social innovations are associated with, and facilitated by, increased sociality (Fox et al. 1999; van Schaik 1999; van Schaik et al. 2003).[16] The flexibility inherent (and observed) in the dispersed sociality of orangutans is dramatic (especially if you include captive colonies). A core aspect of their behavioural response is the fact that in areas of high fruit density/low feeding competition they radically alter their behavioural profiles to engage in much higher rates of affiliation. Also, the flexible use of tools in adaptive extra-somatic foraging also indicates the potential for particularly robust behaviourally flexible inheritances (Malone et al. 2012). In an elegant study, Michael Krützen and colleagues at the University of Zurich compared geographic variation in orangutan behavioural ecology, social organisation and cultural traits. Krützen et al. (2011) determined that salient environmental differences could explain a significant portion of the variation in behavioural ecology and social organisation. Importantly, these differences did not co-vary with genetic variation. Rather, developmental plasticity and social learning create an additional evolutionary pathway whereby locally adaptive phenotypes can be attained. The findings of Krützen et al. (2011: 1809) call into question the widespread tendency to invoke the preeminent action of natural selection in what they term the "de facto null model to explain geographic variation in a trait".

A clearer and more informed picture of behavioural and ecological variation among the three species of orangutan is slowly coming into focus. Given the immediate threats to the remaining populations of these critically endangered apes (e.g., massive deforestation associated with industrial-scale oil palm cultivation), we need to ensure that research efforts inform and shape efficacious conservation action. Are we sure that our models of orangutan socioecology are stimulating the types of studies that will generate applicable datasets? What if our understanding of orangutans is biased by research conducted in habitats where animal density is artificially low (i.e., below carrying capacity)? Comparisons of orangutan density across sites and taxa are fraught with a variety of challenges (e.g., methodological differences; habitat type and fluctuations on both annual and supra-annual cycles); however, some reliable signals can be detected through the noise. First, conservative estimates of orangutan density range between 0.91 and 3.09 individuals/km^2 (Husson et al. 2009). More simply, "orangutans live at population densities that rarely exceed 5 animals/km^2 and are typically below 2.5 animals/km^2" (Meijaard et al. 2010: e12042). Second, orangutan densities do appear to be reliably higher in Sumatra than in Borneo, primarily due to the presence of more consistently productive and favourable habitats (Marshall et al. 2009). In Borneo, compelling evidence suggests that present densities are well below historical (and evolutionary) levels (Meijaard et al. 2010). Meijaard and colleagues point to the previously under-recognised role of hunting, rather than habitat loss, as potentially a key downward-driver of orangutan density and overall abundance over the past 150 years. Indeed, one only needs to read passages from Alfred Russel Wallace's (1869: 34) *The Malay Archipelago*, to get a sense for the ease and access in which hunters encountered orangutans:

> Four days afterward [a previous kill on 12th of May 1855] some Dayaks saw another Mias near the same place, and came to tell me. We found it to be a rather large one, very high up on a tall tree. At the second shot it fell rolling over, but almost immediately got up again and began to climb. At a third shot it fell dead. This was a full-grown female, and while preparing to carry it home, we found a young one face downwards in the bog. This little creature was only about a foot long, and had evidently been hanging to its mother when she first fell.[17]

Prevailing socioecological models that link sociality to ecological determinants (e.g., resource competition) implicitly assume habitats of the focal species are at ecological carrying capacity (e.g., Marshall et al. 2009). If this assumption is not met, then our understanding of orangutan social systems (and the evolutionary mechanisms that undergird its emergence and maintenance) is likely incomplete. Higher densities, increased gregariousness and, ultimately, higher rates of innovation and social learning point to robust pathways whereby organisms shape, and are shaped by social and material exchanges. We have assessed the orangutan to be "adapted" for a semi-solitary/dispersed sociality, but perhaps that is the observable iteration under low-density conditions. Where does this variation in capacity emerge, and can it be seen as an adaptive tailoring to, and tolerance of, periodic cycles of demographic flux?

In orangutans, we see locally adaptive variation emerge via the effects of behavioural flexibility and social learning. This evolved capacity may enable orangutans to adjust their behaviour in response to minor ecological and/or demographic shifts, such as those caused by forest clearance and human activities. Indeed, it is now documented that orangutans can persist in the altered, degraded and regenerating landscapes characteristic of the anthropocene (Wich et al. 2008; Spehar et al. 2018). Such adaptive fine-tuning of socioecological parameters in response to present-day, anthropogenic alterations may be reflective of a responsive versatility by ancestral forms to environmental oscillations (van Schaik et al. 2009; Ibrahim et al. 2013). Detailed paleoenvironmental and paleogeographical reconstructions show that the ancestral orangutan populations (i.e., fossil *Pongo*) faced dramatic challenges in the Late Pleistocene, as continuous, relatively aseasonal forests transformed into refugia with prolonged bouts of seasonal resource reductions (Ibrahim et al. 2013). In portions of their geographical range with intense and widespread climatic effects, orangutans ultimately succumbed to extinction (e.g., South China; Java).[18] Other populations persisted, possibly at reduced densities, in more favourable habitats. Might such challenges in the evolutionary past lead to *resilience* to particular, contemporary threats if similar hurdles have previously been overcome? Balmford (1996) and Coope (1995) refer to this process as "extinction filtering" whereby less vulnerable populations (and the variation within) persist due in part to their previous clearance of similar challenges. Extant orangutans may be able to tolerate minor anthropogenic alterations (excluding hunting) due to their past survival through moderate, climate-driven habitat shifts. However, even low levels of hunting pressure, especially in combination with moderate to severe habitat degradation, will inevitably drive these socially flexible apes into extinction.

Next steps

In this chapter, I have attempted to stimulate thought, problematise the existing paradigm and (gently) agitate some established primatological power structures. I reiterate that I am not advocating for a specific prescription, or calling for a singular, theoretical realignment. The dialectical process sees synthesis emerging from the confrontation between thesis and antithesis (the Hegelian triad), or the "melding of [old and new systems] into a novel theory preserving worthy aspects of both" (Gould 2002: 591). The combination of my personal research insights with the more general call to consider the theoretical developments articulated in the EES have produced an enhanced analysis of three exemplary, social phenomena in a subset of the living apes. In the above description of hylobatid communities, the evolution of behavioural differences in the genus *Pan*, and orangutan social flexibility, I demonstrate the value of this system. Importantly, this approach moves beyond criticism and a simple rejection of reductionism. I echo the perspective of Lewontin and Levins (2007: 103):

> But critique is not just criticism, and dialectics goes beyond the rejection of reductionist or idealist thinking to offer a coherent alternative, more for the way in which it poses questions than for the specific answers its advocates have proposed at any particular time. Its focus is on wholeness and inter-penetration, the structure of process more than of things, integrated levels, historicity, and contradiction. All of this is applied to the objects of the study, to the development of thought about those objects, and self-reflexively to the dialecticians ourselves so as not to lose sight of the contingency and historicity of our own grappling with the problems we study.

This project is of course incomplete, and I will continue to challenge my own biases and construct novel, theoretical compatibilities in light of new research insights. Beyond theory, the construction of material compatibilities among ape and human populations is of paramount importance, especially in light of the former's decline as resulting primarily from the latter's expansion. As individuals, or collectively, those sympathetic to my positions are encouraged to: (1) generate predictions and hypotheses, derived from a dialectical, extended evolutionary model, that expand on, augment, or at times replace those generated by SET; (2) collect inter-generational data on *inclusive inheritances* comprising genetic, behavioural and ecological data; (3) expand this perspective beyond the apes to test the model's robusticity; and (4) balance research with conservation priorities. Suffice to say, there is much work to be done.

The observed patterns of variability within and between hominoid taxa are simultaneously shaped by, and act as shaping factors of, evolutionary processes. I contend that apes have revealed their ability to adjust behaviourally to minor shifts in ecological and demographic conditions via social innovation and learning. However, these same species are also slow-reproducing and therefore vulnerable to rapid environmental alterations, such as those brought about by humans. Over the course of human (and, more generally, hominoid) evolution, our lineage has navigated new ecologies and niches, including a diverse array of plants and animals. These interactions produce reciprocal effects on bodies and behaviour that are at once biological, cultural and political (Fuentes 2020). Germane questions remain. What are the limits of resilience in the face of human activities? How can we use that knowledge to inform conservation theory and practice? In the next chapter, I begin to address these questions by embarking on an in-depth look at a proverbial canary in the coal mine: Java, Indonesia. During my various projects, simultaneous attention to both socio-ecological data and ethnographic insights has revealed the complex ecologies and nuanced realities of primate life in Java. My objectives in Java have always been two-fold: to obtain socioecological data and to contribute to conservation measures. These goals have become integrated and mutually reinforcing over time, and are now bound by theoretical ties that attend to the entanglement of human and other primates' lives in the anthropocene.

Notes

1 In a practical sense, early life peer-association may be an efficacious, yet under-utilised guiding principle in the rehabilitation of displaced apes. Beyond intuition, empirically linking developmental context to adult outcomes remains a challenge (Sherrow & MacKinnon 2011). As a result, researchers often focus on quantifying the strength of adult relationships rather than their ontogeny. For example, in gibbon rehabilitation there is an almost exclusive emphasis on establishing the adult pair bond (see Palmer & Malone 2018), while the efficacy and ethics of orangutan "forest schools", as compared to surrogate mothers (who are often human caregivers) are debated (Russon et al. 2009; Palmer 2020).

2 Gould (1989) spent considerable effort to describe the two distinct meanings of the word *constraint* as it applies to evolutionary biology, noting both the positive (channelling) and negative (restrictive) connotations. In sum, Gould (2002) states that "those who belittle the evolutionary importance of the subject do not deny the phenomenon itself, but rather limit their concept to the negative meaning" (323).

3 According to Roseberry (1989), historical **political economy** is both the "attempt to understand the emergence of particular peoples at the conjunction of local and global histories, to place local populations in the larger currents of world history", and "the attempt to constantly place culture in time, to see a constant interplay between experience and meaning in a context in which both experience and meaning are shaped by inequality and domination" (49). **Political ecology** is

driven by dissatisfaction with dominantly "apolitical" explanations of changing human–environment relationships (Wolf 1972; Nygren & Rikoon 2008).

4 Kappeler and van Schaik (2002: 709–710), in an attempt to reduce the confusion and conflation of terminology, call for the consistent usage of the following definitions: *social organisation* – the size, composition and cohesion of a society in time and space (e.g., pair-living; group-living); *social structure* – the pattern of social interactions and the resulting relationships among the members of a society; and *mating system* – mating interactions and the predicted reproductive consequences (e.g., monogamy; polygyny). While certain linkages are somewhat predictable among the social categories (e.g., individuals in a two-adult group often mate monogamously), researchers should not always assume congruency. Researchers should be specific in which aspects of the social system is being investigated/discussed. Throughout the book I have tried to be consistent and clear with my use of the terms.

5 In *The Structure of Scientific Revolutions*, Kuhn (1970) provides a clear definition of *paradigm* in the context of scientific praxis: "universally recognised scientific achievements that for a time provide model problems and solutions to a community of practitioners" (viii).

6 Kenneth Weiss, Anne Buchanan and Brian Lambert make an important point about the connotations of particular terms. Weiss et al. (2011: 5–6) state that:

> We – especially anthropologists – should realise that we are borrowing cultural terms as if they had no burden of prior connotations. While competition can be defined mathematically in terms of natural selection, it gives misleading connotations to describe selection as "competition" (much less "selfish" or other kindred terms). A green bug camouflaged on a leaf and not eaten can be said in that sense to be competing with a red bug that lands on the leaf and is promptly seen and gulped down by a bird. Mathematically, "green" genes may do better than "red" ones. But the bugs are not competing with each other the way football teams or businesses do. Yet this subtle nuance of intentional or even cognitive competition has seized much of biology, often in a fundamentalistic way.

7 The Japanese tradition of primatology (and the "Kyoto School" in particular) is arguably less constrained by the reductionist paradigms of Western science (Imanishi 1941/2002). Heavily influenced by Kinji Imanishi's perspective – "a sort of holism, in which existence cannot be reduced to its parts" – a distinctive Japanese view of nature is discernible (Matsuzawa & McGrew 2008: 590). Importantly, these views both resonate and underpin theoretical and methodological developments in contemporary primatology (Asquith 2000; Jost Robinson & Remis 2018).

8 For a thorough treatment of hierarchal selection theory (i.e., "group", or interdemic selection; species selection), including a summary of Wilson and Sober's (1994) argument regarding female biased sex-rations, see Gould (2002).

9 Arguably, the clearest demonstration of these dynamics can be seen within the polarising debates surrounding the framing of infanticide (the act of killing an infant) as a sexually selected reproductive strategy. For summaries of the various perspectives, see Bartlett et al. (1993), Sommer (2000) and Rees (2009).

10 It is important to note that there is a degree of push-back from defenders of standard evolutionary theory. For example, see Scott-Phillips et al. (2014) for a debate between "advocates" and "skeptics" of niche construction theory, a core component to the extended evolutionary synthesis. Similarly, Douglas Futuyma (2017)

disagrees that standard evolutionary theory requires any major revision, and instead sees orthodox evolutionary theory as sufficiently robust and flexible to accommodate new developments. Of particular relevance, Futuyama (2017: 1) argues that "the union of population genetic theory with mechanistic understanding of developmental processes enables more complete understanding by joining ultimate and proximate causation; but the latter does not replace or invalidate the former".

11 Jessica Flack and colleagues, based on studies of captive pigtailed macaques (*Macaca nemestrina*), provide evidence for social niche construction in anthropoid primates more generally, where social networks constitute essential social resources within gregarious primate species. Flack et al. (2006: 426) posit that "the structure of such networks plays a critical role in infant survivorship, emergence and spread of cooperative behaviour, social learning and cultural traditions". See also Fuentes (2011) and MacKinnon and Fuentes (2011).

12 Karen Kramer (2019) presents in detail the population history of our species (exceeding 7.7 billion in 2019), and describes a "staggering capacity for population growth" attributed to: 1) a robust repertoire of cooperative strategies; and 2) "evolved life histories favouring shorter birth intervals and higher juvenile survivorship, and within the past two centuries to substantial gains in old age survivorship and reductions in infant mortality" (491). Human growth projections (until 2050) continue to be steep within primate range regions (particularly for Africa, including Madagascar) (Estrada et al. 2017).

13 The following quote by Ryne Palombit (1996: 350) is indicative of the limitations of classically applied descriptors of hylobatid sociality: "the labels typically employed to designate hylobatid societies – 'monogamous' and 'territorial' – may provide a useful context for analysing social behaviour, but they underestimate social variation". A handful of years later, Fuentes (2000: 56) concludes that "given our current data set, it is apparent that the hylobatids are not 'monogamous' primates, although monogamy is a mating pattern that may characterise a number of individuals in a population at any given time", and in doing so destabilised assumptions about a so-called "monogamy package" whereby aspects of social organisation and structure are invariably linked to the mating pattern.

14 Other headlines included: "Chimps and humans are both natural born killers" (John von Radowitz, *Independent*, 17 September 2014); and "Natural born killers: chimpanzees are inherently violent and wage war like human cousins" (Victoria Woollaston, *Daily Mail*, 18 September 2014).

15 A majority of killings (66%) involved inter-community attacks (Wilson et al. 2014). Chimpanzees have been studied for longer, more consistently and in more sites across a broad geographic distribution. The number of bonobo studies, and their cumulative duration, pale in comparison to those of chimpanzees. While increased bonobo research would be expected to reveal further cases of violence or lethal aggression, it is unclear, or perhaps unlikely, if the rate would rise to that of what has been observed in chimpanzees (Sommer et al. 2011). Even so, Wilson and colleagues pooled data from numerous long-term studies for a total of 426 observation years, meaning a killing is observed on average, only once every 7.34 years.

16 Across orangutan taxa and geographic distribution, between 20 and 40 traits (both clear and tentative) attributable to cultural transmission have been identified. These traits include population-specific feeding techniques, social behaviours and communicative signals (van Schaik et al. 2003; van Noordwijk et al. 2006).

17 Wallace chronicled an approximately nine-month stay (March to November 1855) in Sarawak (Malaysian Borneo). Wallace frequently encountered *mias* (or *"mawas"*, another iteration of the local Dayak name for the orangutan). All in all, Wallace shot approximately 30 orangutan, preserving skins and skeletons for museum collections. This particular incident also documents how live-infants can also be procured as a by-product of hunting females. In such a slow-reproducing species, hunting practices such as this would have a devastating impact on both the census and effective population sizes.

18 The extinction of *Pongo* in Java is attributed to dramatic forest loss due to changing climatic conditions in the Late Pleistocene (Storm et al. 2005; Ibrahim et al. 2013).

References

Allman JM, Tetreault NA, Hakeem AY, Manaye KF, Semendeferi K, Erwin JM, Park S, Goubert V, Hof PR. 2010. The von Economo neurons in the frontoinsular and anterior cingulate cortex in great apes and humans. *Brain Structure and Function* 214: 495–517.

Andre C. 2011. Annual Report. *Les Amis des Bonobos du Congo*. Kinshasa, DRC.

Asquith PJ. 2000. Negotiating science: Internalization and Japanese primatology. In Strum SC, Fedigan LM (Eds.), *Primate Encounters: Models of Science, Gender, and Society* (pp. 165–183). Chicago, IL: University of Chicago Press.

Baedke J, Fábregas-Tejeda A, Vergara-Silva F. 2020. Does the extended evolutionary synthesis entail extended explanatory power? *Biology & Philosophy* 35(1): 20.

Balmford A. 1996. Extinction filters and current resilience: The significance of past selection pressures for conservation biology. *Trends in Ecology & Evolution* 11(5): 193–196.

Bartlett TQ. 2009. Seasonal home range use and defendability in white-handed gibbons (*Hylobates lar*) in Khao Yai National Park. In Lappan S, Whittaker DJ. (Eds.), *The Gibbons: New Perspectives on Small Ape Socioecology and Population Biology* (pp. 265–275). New York, NY: Springer.

Bartlett TQ. 2011. The Hylobatidae: Small apes of Asia. In Campbell CJ, Fuentes A, MacKinnon KC, Bearder SK, Stumpf RM. (Eds.), *Primates in Perspective* (2nd ed., pp. 300–312). Oxford: Oxford University Press.

Bartlett TQ, Sussman RW, Cheverud JM. 1993. Infant killing in primates: A review of observed cases with specific reference to the sexual selection hypothesis. *American Anthropologist* 95(4): 958–990.

Behie AM, Teichroeb JA, Malone N. (Eds.). 2019. *Primate Research and Conservation in the Anthropocene*. Cambridge: Cambridge University Press.

Blaikie P, Brookfield H. (Eds.). 1987. *Land Degradation and Society*. London: Methuen & Co., Ltd.

Boose K, White F. 2017. Harassment of adults by immatures in bonobos (*Pan paniscus*): Testing the exploratory aggression and rank improvement hypotheses. *Primates* 58(4): 493–504.

Campbell-Smith G, Campbell-Smith M, Singleton I, Linkie M. 2011. Apes in space: Saving an imperilled orangutan population in Sumatra. *PLoS ONE* 6(2): e17210.

Clutton-Brock TH. 1974. Primate social organization and ecology. *Nature* 250: 539–542.

Clutton-Brock TH, Harvey PH. 1977. Primate ecology and social organization. *Journal of Zoology* 183(1): 1–39.

Clutton-Brock T, Janson C. 2012. Primate socioecology at the crossroads: Past, present, and future. *Evolutionary Anthropology: Issues, News, and Reviews* 21(4): 136–150.

Coope GR. 1995. Insect faunas in ice age environments: Why so little extinction. In Lawton JH, May RM. (Eds.), *Extinction Rates* (pp. 55–74). Oxford: Oxford University Press.

Di Fiore A, Rendall D. 1994. Evolution of social organization: A reappraisal for primates by using phylogenetic methods. *Proceedings of the National Academy of Sciences* 91(21): 9941–9945.

Dingemanse NJ, Kazem AJ, Réale D, Wright J. 2010. Behavioural reaction norms: Animal personality meets individual plasticity. *Trends in Ecology & Evolution* 25(2): 81–89.

Dore KM. 2018. Ethnoprimatology without conservation: The political ecology of farmer-green monkey (*Chlorocebus sabaeus*) relations in St Kitts, West Indies. *International Journal of Primatology* 39(5): 918–944.

Ellwanger AL, Lambert JE. 2018. Investigating niche construction in dynamic human-animal landscapes: Bridging ecological and evolutionary timescales. *International Journal of Primatology* 39(5): 797–816.

Engh AL, Beehner JC, Bergman TJ, Whitten PL, Hoffmeier RR, Seyfarth RM, Cheney DL. 2006. Female hierarchy instability, male immigration and infanticide increase glucocorticoid levels in female chacma baboons. *Animal Behaviour* 71(5): 1227–1237.

Escobar A. 1999. Steps to an antiessentialist political ecology. *Current Anthropology* 40: 1–30.

Estrada A, Garber PA, Rylands AB, Roos C, Fernandez-Duque E, Di Fiore A, … & Rovero F. 2017. Impending extinction crisis of the world's primates: Why primates matter. *Science Advances* 3(1): e1600946.

Flack JC, Girvan M, de Waal FBM, Krakauer DC. 2006. Policing stabilizes construction of social niches in primates. *Nature* 439: 426–429.

Foley R, Lee PC. 1989. Finite social space, evolutionary pathways, and reconstructing hominid evolution. *Science* 243: 901–906.

Fox EA, Sitompul AF, Van Schaik CP. 1999. Intelligent tool use in wild Sumatran orangutans. In Parker ST, Mitchell RW, Miles HL. (Eds.), *The Mentality of Gorillas and Orangutans* (Vol. 480, pp. 99–116). Cambridge: Cambridge University Press.

Fry DP. 2013. *War, Peace and Human Nature: The Convergence of Evolutionary and Cultural Views*. Oxford: Oxford University Press.

Fuentes A. 2000. Hylobatid communities: Changing views on pair bonding and social organization in hominoids. *Yearbook of Physical Anthropology* 43: 33–60.

Fuentes A. 2010. Naturalcultural encounters in Bali: Monkeys, temples, tourists, and ethnoprimatology. *Cultural Anthropology* 25: 600–624.

Fuentes A. 2011. Social systems and socioecology: Understanding the evolution of primate behaviour. In Campbell C, Fuentes A, MacKinnon K, Bearder S, Stumpf R. (Eds.), *Primates in Perspective* (2nd ed., pp. 500–511). Oxford: Oxford University Press.

Fuentes A. 2020. Biological anthropology's critical engagement with genomics, evolution, race/racism, and ourselves: Opportunities and challenges to making a difference in the academy and the world. *American Journal of Physical Anthropology* 175(2): 326–338.

Fuentes A, Baynes-Rock M. 2017. Anthropogenic landscapes, human action and the process of co-construction with other species: Making anthromes in the Anthropocene. *Land* 6(1): 15.

Furuichi T. 2011. Female contributions to the peaceful nature of bonobo society. *Evolutionary Anthropology: Issues, News, and Reviews* 20(4): 131–142.

Futuyma DJ. 2017. Evolutionary biology today and the call for an extended synthesis. *Interface Focus* 7(5): 20160145.

Goodman AH. 2013. Bringing culture into human biology and biology back into anthropology. *American Anthropologist* 115(3): 359–373.

Gould SJ. 1989. A developmental constraint in Cerion, with comments on the definition and interpretation of constraint in evolution. *Evolution* 43(3): 516–539.

Gould SJ. 1992. Shades of Lamarck. In Gould SJ. (Ed.), *The Panda's Thumb: More Reflections in Natural History* (pp. 76–84). New York, NY: W.W. Norton & Company.

Gould SJ. 2002. *The Structure of Evolutionary Theory.* Cambridge, MA: Harvard University Press.

Gould SJ, Lewontin RC. 1979. The spandrels of San Marco and the Panglossian paradigm: A critique of the adaptationist programme. *Proceedings of the Royal Society of London. Series B. Biological Sciences* 205(1161): 581–598.

Hare B, Wobber V, Wrangham R. 2012. The self-domestication hypothesis: Evolution of bonobo psychology is due to selection against aggression. *Animal Behaviour* 83(3): 573–585.

Harrison T. 1996. The palaeoecological context at Niah Cave, Sarawak: Evidence from the primate fauna. *Bulletin of the Indo-Pacific Prehistory Association* 14: 90–100.

Hockings KJ, McLennan MR. 2012. From forest to farm: Systematic review of cultivar feeding by chimpanzees – management implications for wildlife in anthropogenic landscapes. *PLoS ONE* 7(4): e33391.

Hockings KJ, McLennan MR, Carvalho S, Ancrenaz M, Bobe R, Byrne RW, Dunbar RI, Matsuzawa T, McGrew WC, Williamson EA, Wilson ML. 2015. Apes in the Anthropocene: Flexibility and survival. *Trends in Ecology & Evolution* 30(4): 215–222.

Hohmann G, Fruth B. 2011. Is blood thicker than water? In Robbins M, Boesch C. (Eds.), *Among African Apes: Stories and Photos from the Field* (pp. 61–76). Berkeley: University of California Press.

Husson SJ, Wich SA, Marshall AJ, Dennis RD, Ancrenaz M, Brassey R, … & Singleton I. 2009. Orangutan distribution, density, abundance and impacts of disturbance. In Wich SA, Setia TM, van Schaik CP. (Eds.), *Orangutans: Geographic Variation in Behavioral Ecology and Conservation* (pp. 77–96). Oxford: Oxford University Press.

Ibrahim YK, Tshen LT, Westaway KE, Cranbrook E, Humphrey L, Muhammad RF, … & Peng LC. 2013. First discovery of Pleistocene orangutan (*Pongo* sp.) fossils in Peninsular Malaysia: Biogeographic and paleoenvironmental implications. *Journal of Human Evolution* 65(6): 770–797.

Imanishi K. 1941. *A Japanese View of Nature: The World of Living Things.* Translated in 2002 by Asquith PJ, Kawakatsu H, Yagi S, Takasaki H. New York, NY: Routledge Curzon.

Jablonka E, Lamb MJ. 2008. Soft inheritance: Challenging the modern synthesis. *Genetics and Molecular Biology* 31(2): 389–395.

Jablonka E, Lamb M. 2014. *Evolution in Four Dimensions: Genetic, Epigenetic, Behavioural, and Symbolic Variation in the History of Life*, 2nd ed. Cambridge, MA: MIT Press.

Jaeggi AV, Boose KJ, White FJ, Gurven M. 2016. Obstacles and catalysts of cooperation in humans, bonobos, and chimpanzees: Behavioural reaction norms can help explain variation in sex roles, inequality, war and peace. *Behaviour* 153(9–11): 1015–1051.

Jost Robinson CA, Remis MJ. 2018. Engaging holism: Exploring multispecies approaches in ethnoprimatology. *International Journal of Primatology* 39(5): 776–796.

Kamilar JM, Ledogar JA. 2011. Species co-occurrence patterns and dietary resource competition in primates. *American Journal of Physical Anthropology* 144(1): 131–139.

Kano T. 1992. *The Last Ape: Pygmy Chimpanzee Behaviour and Ecology*. Stanford, CA: Stanford University Press.

Kappeler P, van Schaik CP. 2002. Evolution of primate social systems. *International Journal of Primatology* 23(4): 707–740.

Kendal J, Tehrani JJ, Odling-Smee J. 2011. Human niche construction in interdisciplinary focus. *Philosophical Transactions of the Royal Society B* 366: 785–792.

Knott CD, Beaudrot L, Snaith T, White S, Tschauner H, Planansky G. 2008. Female-female competition in Bornean orangutans. *International Journal of Primatology* 29: 975–997.

Koenig A, Borries C. 2009. The lost dream of ecological determinism: Time to say goodbye? … Or a White Queen's proposal? *Evolutionary Anthropology* 18: 166–174.

Kramer KL. 2019. How there got to be so many of us: The evolutionary story of population growth and a life history of cooperation. *Journal of Anthropological Research* 75(4): 472–497.

Krebs JR, Davies NB. (Eds.). 2009. *Behavioural Ecology: An Evolutionary Approach, Fourth Edition*. Hoboken, NJ: John Wiley & Sons.

Krützen M, Willems EP, van Schaik CP. 2011. Culture and geographic variation in orangutan behaviour. *Current Biology* 21(21): 1808–1812.

Kuhn TS. 1970. *The Structure of Scientific Revolutions*, 2nd ed. Chicago, IL: University of Chicago Press.

Laland KN, Odling-Smee J, Feldman MW. 2001. Niche construction, ecological inheritance, and cycles of contingency in evolution. In Oyama S, Griffiths PE, Gray RD. (Eds.), *Cycles of Contingency: Developmental Systems and Evolution* (pp. 117–26). Cambridge, MA: MIT Press.

Laland KN, Uller T, Feldman MW, Sterelny K, Müller GB, Moczek A, Jablonka E, Odling-Smee J. 2015. The extended evolutionary synthesis: Its structure, assumptions and predictions. *Proceedings of the Royal Society B: Biological Sciences* 282(1813): 20151019.

Langergraber KE, Mitani JC, Vigilant L. 2007. The limited impact of kinship on cooperation in wild chimpanzees. *Proceedings of the National Academy of Sciences* 104(19): 7786–7790.

Lappan S, Malaivijitnond S, Radhakrishna S, Riley EP, Ruppert N. 2020. The human–primate interface in the New Normal: Challenges and opportunities for primatologists in the COVID-19 era and beyond. *American Journal of Primatology* 82: e23176.

Lappan S, Whittaker DJ. (Eds.). 2009. *The Gibbons: New Perspectives on Small Ape Socioecology and Population Biology*. New York, NY: Springer.

Levins R, Lewontin R. 1985. *The Dialectical Biologist*. Cambridge, MA: Harvard University Press.

Lewontin R, Levins R. 2007. *Biology Under the Influence: Dialectical Essays on Ecology, Agriculture and Health*. New York, NY: Monthly Review Press.

Longo SB, Malone N. 2006. Meat, medicine, and materialism: A dialectical analysis of human relationships to nonhuman animals and nature. *Human Ecology Review* 13: 111–121.

MacKinnon KC, Fuentes A. 2011. Primates, niche construction, and social complexity: The roles of social cooperation and altruism. In Sussman RW, Cloninger RC. (Eds.), *Origins of Altruism and Cooperation* (Vol. 36, Part 2, pp. 121–143). Developments in Primatology: Progress and Prospects. New York, NY: Springer.

Malone N. 2007. *The Socioecology of the Critically Endangered Javan Gibbon (Hylobates moloch): Assessing the Impact of Anthropogenic Disturbance on Primate Social Systems*. PhD Dissertation, University of Oregon, Eugene, OR.

Malone N, Fuentes A. 2009. The ecology and evolution of hylobatid communities: Proximate and ultimate considerations of inter- and intraspecific variation. In Whittaker D, Lappan S. (Eds.), *The Gibbons: New Perspectives on Small Ape Socioecology and Population Biology* (Chp. 12, pp. 241–264). Series Title: Developments in Primatology: Progress and Prospects. New York, NY: Springer Academic Press.

Malone N, Fuentes A, White FJ. 2012. Variation in the social systems of extant hominoids: Comparative insight into the social behavior of early hominins. *International Journal of Primatology* 33(6): 1251–1277.

Malone N, Oktavinalis H. 2006. The socio-ecology of the silvery gibbon (*Hylobates moloch*) in the Cagar Alam Leuweung Sancang (CALS), West Java, Indonesia. *American Journal of Physical Anthropology* 42(Suppl. 1): 124.

Malone N, Wedana M. 2019. Struggling for socio-ecological resilience: A long-term study of silvery gibbons (*Hylobates moloch*) in the fragmented Sancang Forest Nature Reserve, West Java, Indonesia. In Behie A, Teichroeb J, Malone N (Eds.), *Primate Research and Conservation in the Anthropocene* (pp. 17–32). Cambridge: Cambridge University Press.

Marks J. 2009. *Why I am Not a Scientist: Anthropology and Modern Knowledge*. Berkeley, CA: University of California Press.

Marshall AJ, Ancrenaz M, Brearley FQ, Fredriksson GM, Ghaffar N, Heydon M, … & Proctor J. 2009. The effects of forest phenology and floristics on populations of Bornean and Sumatran orangutans. In Wich SA, Setia TM, van Schaik CP. (Eds.), *Orangutans: Geographic Variation in Behavioral Ecology and Conservation* (pp. 97–117). Oxford: Oxford University Press.

Marx K. 1973. *Grundrisse*, translated by M. Nicolaus. New York, NY: Penguin Books.

Matsuzawa T, McGrew WC. 2008. Kinji Imanishi and 60 years of Japanese primatology. *Current Biology* 18(14): 587–591.

Maynard Smith J. 1982. *Evolution and the Theory of Games*. Cambridge: Cambridge University Press.

McKinney T. 2015. A classification system for describing anthropogenic influence on nonhuman primate populations. *American Journal of Primatology* 77: 715–726.

McNamara JM, Houston AI. 2009. Integrating function and mechanism. *Trends in Ecology & Evolution* 24(12): 670–675.

Meijaard E, Welsh A, Ancrenaz M, Wich S, Nijman V, Marshall AJ. 2010. Declining orangutan encounter rates from Wallace to the present suggest the species was once more abundant. *PLoS ONE* 5(8): e12042.

Nowak MA. 2006. Five rules for the evolution of cooperation. *Science* 314(5805): 1560–1563.

Nygren A, Rikoon S. 2008. Political ecology revisited: Integration of politics and ecology does matter. *Society and Natural Resources* 21(9): 767–782.

Odling-Smee J. 2007. Niche inheritance: A possible basis for classifying multiple inheritance systems in evolution. *Biological Theory* 2(3): 276–289.

Odling-Smee J, Laland KN, Feldman MW. 2003. *Niche Construction: The Neglected Process in Evolution*. Princeton, NJ: Princeton University Press.

Palmer A. 2020. *Ethical Debates in Orangutan Conservation*. London: Routledge.

Palmer A, Malone N. 2018. Extending ethnoprimatology: Human–alloprimate relationships in managed settings. *International Journal of Primatology* 39(5): 831–851.

Palombit R. 1996. Pair bonds in monogamous apes: A comparison of the siamang (*Hylobates syndactylus*) and the white-handed gibbon (*Hylobates lar*). *Behaviour* 133: 321–356.

Parga JA, Overdorff DJ. 2011. Primate socioecology. In Campbell C, Fuentes A, MacKinnon K, Bearder S, Stumpf R. (Eds.), *Primates in Perspective* (2nd ed., pp. 12–18). Oxford: Oxford University Press.

Piersma T, Drent J. 2003. Phenotypic flexibility and the evolution of organismal design. *Trends in Ecology and Evolution* 18: 228–233.

Ramachandran VS. 2011. *The Tell-Tale Brain: A Neuroscientist's Quest for What Makes us Human*. New York, NY: W.W. Norton & Company.

Rappaport RA. 1968. *Pigs for the Ancestors*. New Haven, CT: Yale University Press.

Rees A. 2009. *The Infanticide Controversy: Primatology and the Art of Field Science*. Chicago, IL: University of Chicago Press.

Reichard UH. 2009. The social organisation and mating systems of Khao Yai white-handed gibbons: 1992–2006. In Lappan S, Whittaker DJ. (Eds.), *The Gibbons: New Perspectives on Small Ape Socioecology and Population Biology* (pp. 347–384). New York, NY: Springer.

Reichard UH, Barelli C, Hirai H, Nowak MG. 2016. The evolution of gibbons and siamang. In Reichard U, Hirai H, Barelli C. (Eds.), *Evolution of Gibbons and Siamang* (pp. 3–41). Developments in Primatology: Progress and Prospects. New York, NY: Springer.

Reichard UH, Ganpanakngan M, Barelli C. 2012. White-handed gibbons of Khao Yai: Social flexibility, complex reproductive strategies, and a slow life history. In Kappeler P, Watts D. (Eds.), *Long-Term Field Studies of Primates* (pp. 237–258). Berlin, Heidelberg: Springer.

Reisland MA. 2013. *Conservation in a Sacred Forest: An Integrated Approach to Assessing the Management of a Community-Based Conservation Site*. PhD Thesis, University of Wisconsin, Madison.

Reisland MA, Lambert JE. 2016. Sympatric apes in sacred forests: Shared space and habitat use by humans and endangered Javan gibbons (*Hylobates moloch*). *PLoS ONE* 11(1): e0146891.

Riley EP. 2020. *The Promise of Contemporary Primatology*. New York, NY: Routledge.

Riley EP, Fuentes A. 2010. Conserving social–ecological systems in Indonesia: Human-nonhuman primate interconnections in Bali and Sulawesi. *American Journal of Primatology* 73(1): 62–74.

Robbins D, Chapman CA, Wrangham RW. 1991. Group size and stability: Why do gibbons and spider monkeys differ? *Primates* 32(3): 301–305.

Roseberry W. 1989. *Anthropologies and Histories*. New Brunswick, NJ: Rutgers University.

Roth TS, Rianti P, Fredriksson GM, Wich SA, Nowak MG. 2020. Grouping behavior of Sumatran orangutans (*Pongo abelii*) and Tapanuli orangutans (*Pongo tapanuliensis*) living in forest with low fruit abundance. *American Journal of Primatology* 82(5): e23123.

Russon AE, Adams L, Kuncoro P, Smith J. 2009. *Designing Forest Schools for the Behavioural Rehabilitation of Immature Ex-Captive Orangutans*. Toronto: York University & BOS Canada.

Sahlins M. 1972. *Stone Age Economics*. New York, NY: de Gruyter.

Scott-Phillips TC, Laland KN, Shuker DM, Dickins TE, West SA. 2014. The niche construction perspective: A critical appraisal. *Evolution* 68(5): 1231–1243.

Sherrow HM, MacKinnon KC. 2011. Juvenile and adolescent primates: The application of life history theory. In Campbell C, Fuentes A, MacKinnon K, Bearder S, Stumpf R. (Eds.), *Primates in Perspective* (2nd ed., pp. 455–464). Oxford: Oxford University Press.

Shipman P, Bosler W, Davis KL, Behrensmeyer AK, Dunbar RIM, Groves CP, ... & Stucky RK. 1981. Butchering of giant geladas at an Acheulian site [and Comments and Reply]. *Current Anthropology* 22(3): 257–268.

Sommer V. 2000. The holy wars about infanticide. Which side are you on? And why? In van Schaik CP, Janson CH (Eds.), *Infanticide by Males and its Implications* (pp. 9–26). Cambridge: Cambridge University Press.

Sommer V, Bauer J, Fowler A, Ortmann S. 2011. Patriarchal chimpanzees, matriarchal bonobos: Potential ecological causes of a *Pan* dichotomy. In Sommer V, Ross C. (Eds.), *Primates of Gashaka* (pp. 469–501). New York, NY: Springer.

Spehar SN, Sheil D, Harrison T, Louys J, Ancrenaz M, Marshall AJ, ... & Meijaard E. 2018. Orangutans venture out of the rainforest and into the Anthropocene. *Science Advances* 4(6): e1701422.

Sterck EHM, Watts DP, van Schaik CP. 1997. The evolution of female social relationships in nonhuman primates. *Behavioural Ecology and Sociobiology* 41: 291–309.

Storm P, Aziz F, de Vos J, Kosasih D, Baskoro S, van den Hoek Ostende LW. 2005. Late Pleistocene *Homo sapiens* in a tropical rainforest fauna in East Java. *Journal of Human Evolution* 49(4): 536–545.

Strier KB. 1994. Myth of the typical primate. *American Journal of Physical Anthropology* 37(S19): 233–271.

Strier KB. 2003. Primatology comes of age: 2002 AAPA luncheon address. *American Journal of Physical Anthropology* 122(S37): 2–13.

Strier KB. 2017a. *Primate Behavioural Ecology*, 5th ed. London: Routledge.

Strier KB. 2017b. What does variation in primate behaviour mean? *American Journal of Physical Anthropology* 162: 4–14.

Struhsaker TT. 1997. *Ecology of an African Rainforest: Logging in Kibale and the Conflict Between Conservation and Exploitation*. Gainesville: University of Florida Press.

Struhsaker TT. 1999. Primate communities in Africa: The consequences of long-term evolution or the artifact of recent hunting? In Fleagle JG, Janson C, Reed KE. (Eds.), *Primate Communities* (pp. 289–294). Cambridge: Cambridge University Press.

Strum SC. 2010. The development of primate raiding: Implications for management and conservation. *International Journal of Primatology* 31: 133–156.

Strum SC. 2012. Darwin's monkey: Why baboons can't become human. *American Journal of Physical Anthropology* 149(S55): 3–23.

Stumpf RM. 2017. Chimpanzee and bonobo (*Pan*). *The International Encyclopedia of Primatology* 1: 1–3.

Sueur C, Romano V, Sosa S, Puga-Gonzalez I. 2019. Mechanisms of network evolution: A focus on socioecological factors, intermediary mechanisms, and selection pressures. *Primates* 60(3): 167–181.

Sussman RW. 2011. A brief history of primate field studies. In Campbell CJ, Fuentes A, MacKinnon KC, Bearder SK, Stumpf RM. (Eds.), *Primates in Perspective* (2nd ed., pp. 6–11). Oxford: Oxford University Press.

Sussman RW, Garber PA, Cheverud JM. 2005. Importance of cooperation and affiliation in the evolution of primate sociality. *American Journal of Physical Anthropology* 128(1): 84–97.

Thierry B. 2008. Primate socioecology, the lost dream of ecological determinism. *Evolutionary Anthropology: Issues, News, and Reviews: Issues, News, and Reviews* 17(2): 93–96.

Thompson M, Warburton M. 1985. Uncertainty on a Himalayan scale. *Mountain Research and Development* 5(2): 115–135.

Tokuyama N, Sakamaki T, Furuichi T. 2019. Inter-group aggressive interaction patterns indicate male mate defense and female cooperation across bonobo groups at Wamba, Democratic Republic of the Congo. *American Journal of Physical Anthropology* 170(4): 535–550.

Tutin CEG, Oslisly R. 1995. *Homo, Pan* and *Gorilla*: Co-existence over 60,000 years at Lope in Central Gabon. *Journal of Human Evolution* 28: 597–602.

Utami SS, Wich SA, Sterck EHM, van Hooff JARAM. 1997. Food competition between wild orangutans in large fig trees. *International Journal of Primatology* 18: 909–927.

van Noordwijk MA, van Schaik CP, Wich SA. 2006. Innovation in wild Bornean orangutans (*Pongo pygmaeus wurmbii*). *Behaviour* 143(7): 839–876.

van Schaik CP. 1999. The socioecology of fission-fusion sociality in orangutans. *Primates* 40: 69–86.

van Schaik CP, Ancrenaz M, Borgen G, Galdikas B, Knott C, Singleton I, Suzuki A, Utami S, Merrill M. 2003. Orangutan cultures and the evolution of material culture. *Science* 299: 102–105.

van Schaik CP, Janson CH. (Eds.). 2000. *Infanticide by Males and its Implications*. New York, NY: Cambridge University Press.

van Schaik CP, Marshall AJ, Wich SA. 2009. Geographic variation in orangutan behavior and biology. In Wich SA, Utami Atmoko SS, Setia TM, van Schaik CP. (Eds.), *Orangutans: Geographic Variation in Behavioural Ecology and Conservation* (pp. 351–361). Oxford: Oxford University Press.

van Schaik CP, van Hooff JARAM. 1983. On the ultimate causes of primate social systems. *Behaviour* 85: 91–117.

Wallace AR. 1869. *The Malay Archipelago*. Singapore: Periplus Editions.

Weiss KM, Buchanan AV, Lambert BW. 2011. The Red Queen and her king: Cooperation at all levels of life. *American Journal of Physical Anthropology* 146(S53): 3–18.

West-Eberhardt MJ. 2003. *Developmental Plasticity*. Oxford: Oxford University Press.

White FJ. 1996. *Pan paniscus* 1973 to 1996: Twenty-three years of field research. *Evolutionary Anthropology: Issues, News, and Reviews: Issues, News, and Reviews* 5(1): 11–17.

Wich SA, Meijaard E, Marshall AJ, Husson S, Ancrenaz M, Lacy RC, ... & Doughty M. 2008. Distribution and conservation status of the orang-utan (*Pongo* spp.) on Borneo and Sumatra: How many remain? *Oryx* 42(3): 329–339.

Williams GC. 1966. Natural selection, the costs of reproduction, and a refinement of Lack's principle. *The American Naturalist* 100(916): 687–690.

Wilson DS, Sober E. 1994. Reintroducing group selection to the human behavioral sciences. *Behavioral and Brain Sciences* 17(4): 585–608.

Wilson ML, Boesch C, Fruth B, Furuichi T, Gilby IC, Hashimoto C, ... & Lloyd, JN. 2014. Lethal aggression in *Pan* is better explained by adaptive strategies than human impacts. *Nature* 513(7518): 414–417.

Wolf E. 1972. Ownership and political ecology. *Anthropological Quarterly* 45(3): 201–205.

Wrangham RW. 1979. On the evolution of ape social systems. *Social Science Information* 18: 335–369.

Wrangham RW. 1980. An ecological model of female-bonded primate groups. *Behaviour* 75: 262–300.

Wrangham RW. 1986. Ecology and social relationships in two species of chimpanzee. In Rubenstein DI, Wrangahm RW. (Eds.), *Ecological Aspects of Social Evolution: Birds and Mammals* (pp. 352–378). Princeton, NJ: Princeton University Press.

Wrangham RW, Pilbeam D. 2001. African apes as time machines. In Galdikas B, Briggs N, Sheeran L, Shapiro G, Goodall J. (Eds.), *All Apes Great and Small* (pp. 5–17). New York, NY: Plenum.

Wrangham RW, White FJ. 1988. Feeding competition and patch size in the chimpanzee species *Pan paniscus* and *Pan troglodytes*. *Behaviour* 105(1–2): 148–164.

York R, Longo SB. 2017. Animals in the world: A materialist approach to sociological animal studies. *Journal of Sociology* 53(1): 32–46.

Zeder MA. 2017. Domestication as a model system for the extended evolutionary synthesis. *Interface Focus* 7(5): 20160133.

4 Waves of change

Insights from Java, Indonesia

An ocean without its unnamed monsters would be like a completely dreamless sleep.

(John Steinbeck)

The banteng (*Bos javanicus*), or wild forest cattle, was meant to be locally extinct within the Sancang Forest Nature Reserve. Yet one appeared to me, at midnight, as I camped within the spiritual heart of this revered forest on the south coast of West Java. Stirred by the sound of clumsy hooves on stone cobbles, I peered towards the river's edge. The moon illuminated a dark, massive frame atop distinctive, white-stockinged legs. We both froze for a moment until the rustling of my tent triggered the rustling of vegetation as the beast retreated. Was this individual the last of its kind in this threatened landscape? Was it an apparition? In this forest, where *karuhun* (ancestors) are routinely conjured by spiritual intermediaries called *kuncen*, perceptions and possibilities are expanded. Still other spirits will remain nameless here, as per their penchant for taking human souls. Over the course of several years at Sancang, I focused my attention on the endangered silvery gibbon (*Hylobates moloch*), and in particular the three groups that ranged directly above some of the most sacred sites (*keramat*) in the forest. These sites are the object of frequent pilgrimage by those seeking to connect with the supernatural powers of the spiritual realm. The gibbons took a variable degree of notice to the mix of locals, researchers and pilgrims below. It was in this context that I set out to understand the behaviour and ecology of this vanishing ape, and to test basic socioecological hypotheses about the emergence of their distinctive social system. However, I was busy filling my notebooks with experiences, stories and myths that seemed at first to be tangential, but would later play a significant role in shaping my thoughts about the efficacy of primate research and conservation activities in Java.

Organising and implementing projects in Indonesia requires considerable patience and perhaps a bit of good fortune. A person can have control of the former, but is of course subject to the fickleness of the latter. Interestingly, many of the spiritual tourists I have met at sacred sites in Java expressed the desire to change their luck. Escaping the dynamics of a bad relationship, breaking a run

DOI: 10.4324/9780367211349-4

Figure 4.1 Kang Heri's 1960s–era Land Rover on one of many trips over the precarious slopes of Gunung Gelap (which translates to "dark mountain") between Bandung and Pameungpeuk, circa 2005. Within moments, a sufficient crowd had gathered to assist in righting the ship. With a turn of the engine crank we were back on our way.

Photo credit: Nicholas Malone.

of unsuccessful financial decisions or a change of fortune in matters related to health are all reasons that have been shared with me as an underlying motivation for their pilgrimage. While I have personally experienced my fair share of "fortunate" moments in Java (Figure 4.1), I have had countless opportunities to develop my proficiency in remaining patient.[1] While it is true that many such opportunities are induced by Indonesia's legendary bureaucratic structures (arguably designed to test the resolve and composure of foreign researchers), patience is also a requisite component of anthropological research, generally speaking. In the Indonesian archipelago specifically, mysteries woven into the sheer complexity of the biogeography, cultural and linguistic diversity, colonial histories and ecologies can be slowly untangled through a combination of patience and persistence. As both a primate biodiversity hotspot and an engine of anthropogenic impact, insights from Java are particularly salient to global conversations about primate research and conservation. What will our relationship with primates (and primatology itself) look like in 10 years? In 100? In this chapter, I aim to draw upon lessons from roughly 20 years of research experience in Java in an effort to identify future challenges and applicable solutions. I do this through the lens of a dialectical primatologist.

Java, Indonesia: Beyond singular description

The biogeographic context

The nation of Indonesia is the world's largest archipelago and encompasses over 13,000 islands. This insular nation is spread over 5120 km, a distance equal to nearly one-eighth of the earth's entire circumference. The Indonesian islands of Sumatra, Borneo, Java and Bali, along with the Thai–Malay peninsula south of the Isthmus of Kra, are collectively known as Sundaland (Nijman 2001). During much of the Pleistocene/Holocene transitional period (~50,000–5,000 years ago), the Sunda Shelf was more fully exposed due to lower sea levels (up to 150 metres lower than present day), and therefore creates opportunities to understand the migration and succession of fauna and flora. Identified as a bio-diversity hotspot, this region contains a large number of endemic plants and vertebrate animals, including 1,500 of the world's 9,000 known bird species (Stark 1989; Myers et al. 2000). Indonesia is a hotspot for primate diversity as well. Although occurring naturally in 90 countries, two-thirds of all living pri-mate species ($n = 504$) occur in just four nations: Indonesia, Brazil, Madagascar and the Democratic Republic of the Congo (DRC) (Estrada et al. 2017). Today, despite the reduced exposure of the Sunda Shelf, Indonesia ranks third in total forested area (behind Brazil and the DRC). This represents approximately one-tenth of the world's remaining tropical rainforest.

The theory of island biogeography posits that the number of species found on an island is determined by the isolation distance and overall size of the island (MacArthur & Wilson 1967). In 2001, an analysis of the Sunda region by Alexander Harcourt and Mark Schwartz substantiated tenets of the island bio-geographic model in that "fewer taxa, including primates, are found on small islands than on large islands, as expected if the insularisation was accompanied by extinction" (5).[2] Moreover, certain biological characteristics are associated with the absence of nonhuman primate species on islands of various sizes following insularisation, including: large body size, low population density, equatorial confinement and large annual ranges (Harcourt & Schwartz 2001). "Islands" can also emerge via processes of habitat fragmentation. The large-scale environmental change during the Pleistocene and Holocene provides the conditions necessary to examine the relationships between organismal response and susceptibility to regional extinction. Nina Jablonski (1998) compared the spatial distributions of five catarrhine primate genera (three hominoid and two cercopithecoid) during dramatic increases in environmental seasonality and the potential for physical isolation due to habitat fragmentation.[3] Jablonski tested the hypothesis that individual genera exhibit differential responses to conditions of increasing seasonality due to variable reproductive, dietary and developmental constraints. The distributions of all five genera shift towards the equator from Early to Late Pleistocene. Within the Hominoidea, large body size is negatively correlated with frequency of occurrence in the increasingly seasonal subtropical zone. Within the Cercopithecoidea, distributions generally extend into higher latitudes than those of the apes for all time periods, and tropical compression is less pronounced for monkeys than for apes during the late Pleistocene and Holocene. Large body size among the apes, in combination with life history variables, greater absolute energetic requirements and the nutritional demands of a large brain, does not equal success in seasonal environments. The gibbons relatively small body size (~5–15 kg) have enabled these ape taxa to persist, albeit scarcely, in moderately seasonal environments. Such correlations have led researchers to the general consensus that island size, rather than human influence (as measured by human density), represents a greater influence on the presence or absence of nonhuman primate species on islands.

Similarly, the extent of deforestation can predict the expected number of taxa threatened with extinction in a given region. However, it is important to delve further into the particular factors driving past losses and future vulnerabilities. For example, on Java fewer endemic bird species are threatened with extinction than predicted by the extent of deforestation (Brooks et al. 1997). Java is also unusual compared to other islands of the Greater Sundas (Sumatra, Borneo and Sulawesi) in possessing a significantly depauperate avifauna at altitudes of 300–1500 metres above sea level (van Balen 1994). Balmford (1996: 194) suggests "that the apparent shortage of currently threatened birds may instead arise because historical forest clearance purged the island of many of its more vulnerable species long ago". On Java, many species have probably been lost to the activity of humans, since more than 100 plant genera found in southern Sumatra do not occur in Java – too many to explain by island biogeography alone (van Steenis & Schippers-Lamerste 1965, as cited in Whitten et al. 1996). Furthermore, Java's nonhuman primate species richness is slightly destitute (five species) compared to either Borneo or mainland Sumatra (minimally 13 species each).

Notwithstanding this varying response by organisms to environmental fluctuations, a more dramatic extinction rate is currently being driven by the domination of ecological systems by humans (Pimm & Brooks 2000). Several noteworthy extinctions of large-bodied, mammalian species have been described from Java. First, although brought to the brink of extinction by dramatic forest losses associated with changing climatic conditions in the Late Pleistocene, the action of human agents likely factor into the local extinction of the orangutan (Storm et al. 2005; Louys et al. 2007; Ibrahim et al. 2013; Chapter 3). Second, the Javan subspecies of tiger (*Panthera tigris sondaica*) is also extinct despite perennial claims of encountering putative signs (e.g., scat, or claw marks on trees). Javan tigers were hunted to extinction for a variety of reasons, including: a) status; b) fear of, or retaliation for, attacks on humans; and c) the individual body parts used in rituals or "medicinal" contexts.[4] Suffice to say, the tiger's extinction can be equated simply with not enough room on the island for both humans and a large, apex predator (Seidensticker 1987). Today, a combination of historical land use patterns, high human density and consistent geophysical forces (e.g., volcanic activity, earthquakes and landslides) transforms Java's surface landscape at a rapid pace. This suite of environmental parameters and anthropogenic pressures forge an engine of ecosystem alteration that, with near-certainty, will lead to further losses of vulnerable taxa.

The cultural, political and economic heart of Indonesia

Austronesian peoples arrived on the island of Java some 3,000–5,000 years ago, and their descendants now dominate western Indonesia (Whitten et al. 1996). Historically, localised agriculture, village life and religious expression (conceived as a historical syncretism of animism with Buddhism, Hinduism and Islam) coalesced into a unifying fibre of social life (Geertz 1960). Buddhist and Hindu influences were prevalent, especially in Central Java, until approximately 1400 AD. Islam was introduced to Indonesia at approximately 1400 AD, and by the end of the 16th century it replaced Buddhism and Hinduism as the dominant religion. In addition to religious influences, the cultivation of rice (*Oryza sativa*) is similarly intertwined with life in the Asian tropics, and, accordingly, rice fields are a ubiquitous feature of Java's landscape. Rice cultivation on Java employs soil and water conservation techniques (terraced, irrigated fields or sawah). By the mid-20th century, the strains of overpopulation were taking a toll on the island's inhabitants, especially on the peasant farmers. This strain was mainly manifested in the form of declining living and nutritional standards, including the daily, per-capita consumption of protein and overall calories. Other indicators of an undernourished populace, such as high rates of infectious disease transmission, were also evident. With its large population, and close to ideal rice producing environment, Java became a prime candidate for the Green Revolution technical package and its high-yield varieties of rice. In short, the successes (and failures) of these initiatives had major impacts on social institutions *and* biodiversity (Franke 1977).

Java's geographic location has made it a historically important crossroads for centuries, with Java's valuable timber being at the centre of economic and political control (Colfer & Resosudarmo 2002; Malone et al. 2014a). Hindus introduced teak trees from 200 AD to 400 AD, and by 1000 AD, there were over 1 million hectares of managed teak, organised into vassal states, in place of native flora (Boomgaard 1988; Koentjaraningrat 1985, as cited in Whitten et al. 1996). The introduction of teak (*Tectona grandis*) and the subsequent domination of this non-indigenous species is primarily responsible for major disruption to forest ecosystems. Teak, coffee, tea and clove plantations, in combination with a growing human population and demand for fuel-wood, resulted in considerable deforestation that consumed both lowland and sub-montane forests up to 1500 metres above sea level. Later, Islamic kingdoms used the teak for shipbuilding, attracting the attention of Holland's fortune-seeking merchants in the East. In 1602 the Dutch government launched the Vereenigde Oost-Indische Compagnie (Dutch East India Company), or VOC. In what Winchester (2004) describes as "a business model for the foundations of all modern capitalism", the VOC exercised "exclusive and quasi-sovereign rights to enter into treaties with local princes, to build forts, to maintain armed forces and to set up administrative systems of governments whose officials pledged loyalty to the government of the Netherlands" (29).

In addition to fuelling the fortunes of its merchants and undergirding Holland's imperial ambitions, the VOC facilitated the Dutch usurpation of Java in order to gain control of teak for wood, or as fuel in the production of coffee, tea, indigo and sugar. After becoming a Dutch colony rather than company territory, the Dutch implemented a cultivation-system, or *cultuurstelsel*, that mandated a percentage of local crop production to be earmarked for government exportation (Whitten et al. 1996). Land that had traditionally been for subsistence crops became coffee, tea, cloves and sugar plantations with significant social and ecological repercussions (Geertz 1963). More and more land that had been worked and directed/managed by the Javanese was converted to plantations, and in turn they cleared more forests for subsistence production. *Cultuurstelsel* was abolished in 1870, but the plantation system had become entrenched, and private companies took over management of the state-run plantations (Taylor 2003). Contrastingly, a state-owned entity (Perhutani) oversees millions of hectares of protection and production forests (which include allowances for some cultivation of forested land). Therefore, as Whitten et al. (1996) pointed out, land that is actually covered with forested habitat does not map precisely on to formalised, forest boundaries.

When Indonesia proclaimed its independence in 1945, the first president, Sukarno, adopted many of the colonial laws and perpetuated external economic control over Indonesia's resources (Dake 2006). His general Suharto led a successful coup against the Sukarno Regime culminating in Sukarno's overthrow in 1967. Suharto's New Order Regime claimed at least 70% of Indonesia's land as political forest and productive agroforests as state resources (McCarthy 2000). In 1998, the Suharto Regime fell. The Reformasi replaced the New Order, and immediately peasants and agrarian farmers reacted to the fall of the Regime by occupying the state forests and monoculture plantations (Applegate et al. 2002). Many landless people chopped down the timber and crops, and replanted their own, and forest fires raged across Indonesia's islands as an expression of frustration (Dudley 2002). Reformasi was meant to signify the changes to the Indonesian economy and the decentralisation of the state. But the historic emphasis on production forests has persisted and foresters continually exceed the recommended logging quotas for conservation, in order to satisfy the quotas for profit (Supriatna et al. 2010). Suffice to say, the particular history of natural resource exploitation on Java reflects a combination of local specificities and extra-local generalities associated with global economic and geopolitical forces, and the results are emblematic of colonial and postcolonial, capitalist penetrations into the global South (Malone et al. 2014a).

Indonesia's formal economy is dominated by the extraction and export of natural resources including oil, gas, rocks/minerals, as well as timber and non-timber forest products (NTFP) (McConkey et al. 2005). This formal sector of wealth extraction (and retention) collapsed amidst the Asian economic crisis of 1997–1998, as the Indonesian stock market crashed (losing 89% of its value) and the value of the Indonesian rupiah reached a temporary low of only 14.7% of its pre-crisis value (Baldwin 2003). A result of the crisis is the transformation and destruction of Indonesia's forests through the dramatic rise in illegal logging and the widespread burning of mature tropical forests. The human population of Java has been transitioning from rural to urban, and concomitantly has become increasingly integrated into the global economy. As these processes often correspond with economic disparities, the continued access to (and ability to exploit) rural and/or forested areas may provide a safety net for some individuals (and their familial support networks) during such transitions. Indeed, the impacts of the economic crises were unequally distributed between urban and rural communities as a function of access to available farm and forest wealth (Wetterberg et al. 1999; Baldwin 2003). In the analysis by Wetterberg et al. (1999), food security and social support in the rural areas of West Java (Propinsi Jawa Barat) were identified as being most effected by the crisis, attributable to the lack of remaining forest and uncultivated rural land. Sundanese (West Javan) people have been shaped by the historical rule by Hindu, Muslim, and Dutch kingdoms, Japanese military oppression, American interventions, dictatorships and more recently by contested forms of democracy hampered by the long reach of centralised power and control (Whitten et al. 1996). Today, as described above, the remaining fragments of primary forest are set among a mosaic of government-controlled protection and production forests, and are continuously altered by both an expanded (and extractive) urban–rural connectivity and those still engaging in subsistence livelihoods (Meijerink 1977; USAID 2004; Malone et al. 2014b). In sum, the lives (and habitats) of forest dwelling primates, and the health and welfare of human communities, are inextricably (and dialectically) intertwined.

The natureculture lens

Of course, many of the above details are highly specific to Indonesia generally, and Java in particular. However, I would argue that researchers could identify similar webs of geographic/political/cultural/ecological variables in all primate range states, and indeed some have (e.g., Oates 1999; Harper 2002; Estrada et al. 2017). This makes lessons from Java potentially generalisable to other contexts. In part, the way systems are framed dictates the interconnections that are discernible. Commonly, conservation policies isolate either the material or cultural components of complex systems, and doing so risks the alienation of some agents from their ecological context. Instead, I will lean into the critical concept of *natureculture*. Natureculture is a synthesis of nature and culture that recognises their inseparability in ecological relationships that are both biophysically and socially formed, and emerges from the scholarly interrogation of dualisms that are deeply embedded within the intellectual traditions of the sciences and humanities (e.g., human/animal; nature/culture) (Haraway 2003; Fuentes 2010; Malone & Ovenden 2017). Åsa Johansson (2015: 17) concisely describes Haraway's conceptualisation of a world that

> continuously co-creates itself through a dynamically changing intrinsic intertwinement of matter, discourse and semiotics. This means that what is commonly referred to as a pure and static nature is just as context-dependent as culture; corporeality and nature are – and always have been – influencing and influenced by historically and locally specific power-loaded sociocultural structures.

As such, natureculture is an effective lens through which to examine the inter-play of biophysical and social elements of complex systems – such as is found in Java.

Alfred Russel Wallace, who travelled extensively throughout insular Southeast Asia, sets the stage for this examination as well as anyone. Wallace (1869: 76) opined:

> Taking it as a whole, and surveying it from every point of view, Java is probably the very finest and most interesting tropical island in the world … it is undoubtedly the most fertile, the most productive, and the most populous island within the tropics. Its whole surface is magnificently varied with mountain and forest scenery. It possesses thirty-eight volcanic moun-tains, several of which rise to ten or twelve thousand feet high. Some of these are in constant activity, and one or other of them displays almost every phenomenon produced by the action of subterranean fires, except regular lava streams, which never occur in Java. The abundant moisture and tropical heat of the climate causes these mountains to be clothed with lux-uriant vegetation, often to their very summits, while forests and plantations cover their lower slopes … Java too possesses a civilisation, a history and antiquities of its own, of great interest … for, scattered through the country, especially in the eastern part of it, are found buried in lofty forests, temples, tombs and statues of great beauty and grandeur; and the remains of exten-sive cities, where the tigers, the rhinoceros, and the wild bull now roam undisturbed.[5]

Wallace's observations affirm that ecologies and geophysical forces, as well as human histories, actions and beliefs all contribute to the mutual shaping of social and ecological worlds. Given the complex and contingent cultural and ecological webs, theories that fail to account for reciprocal, non-linear and interactive process (e.g., positivist/deterministic theories such as cul-tural ecology) are woefully inadequate. This point is particularly relevant as we, at present, continue to co-construct globalised, ecological relationships with other primates (Malone & Ovenden 2017).[6] Natureculture's visibility in the primatological literature parallels the emergence and proliferation of studies in the field of "ethnoprimatology" (Riley 2020). By drawing upon theory and methods from ecological, biological, ethnographic and historical approaches, ethnoprimatology creates a more robust and accurate framework for anthropologists and primatologists interested in understanding the com-plexity of human–alloprimate interfaces in the anthropocene (Sponsel 1997; Wheately 1999; Fuentes & Wolfe 2002; Riley 2005, 2007; Malone et al. 2014b; Dore et al. 2017). Using the ethnoprimatological framework, in combination with the natureculture critique, offers the potential for new insights into multi-layered, socioecological relationships. Here, I build upon previous work to argue that attention to these entanglements not only provides an exciting arena for the diverse theoretical and methodological toolkits of anthropology and primatology to interact (Fuentes 2012; Riley 2020), but also has the potential to produce timely insights into the facilitation of species-coexistence.

The sacred forests of Sunda

The Dutch in Batavia (Jakarta) created a material, fortified boundary for their settlement out of fear for the unfamiliar and expansive jungle. Winchester (2004: 41), with respect to the fortification, wrote:

> Beyond it stretched the jungles, hot, dense, soggy, ever hostile, and alive with animals: the tiger and the panther, the tapir and the one-horned rhino, black apes [presumably Javan langurs, or lutung] and giant rats, a range of giant pythons and venomous cobras together with a gaudy wealth of cockatoos, parrots, and birds of paradise.

Boundaries can also be conceptual, such as those between natural and supernatural realms (e.g., relationships between living descendants and ancestral spirits), or temporal boundaries that demarcate historical eras (Wessing 1993).[7] For Sundanese (West Javan) people, in stark contrast to the Dutch, those same forests (and their denizens) are conceived of as powerful mediating entities that guard, and therefore preserve the boundaries between nature and society, broadly glossed. Robert Wessing, a cultural anthropologist at Leiden University, has written extensively about symbolic and social life in Java. Drawing on the work of both eastern and western scholars (e.g., Ajip Rosidi and Reimer Schefold), as well as his own extensive research, Wessing offers a clearly articulated synopsis of the nuanced role of forests in the Sundanese worldview. Wessing (1993: 5) states that

> nature and especially the forest should not be seen as an unordered, chaotic entity opposed to the order of society. It has its own rules, similar to but different from those of human society. However, whereas society is controlled by knowable human rules, those of the forest are less apparent and thus the forest is dangerous, precisely because one does not know the rules.

Enter shamans and tigers. Belief systems pre-dating Hinduism in Java envisaged the manifestation of divinities embodying earth's fertility, or "earth gods", in crops, livestock and forests. A communicative bridge between the earthly realm and the realm of the divine can be facilitated by a shaman – one capable of navigating and becoming "one with the forest" – possessing the ability to safely interact with the powers situated within the forest (Wessing 1993). Likewise, tigers, once relatively common occupants of the habitat margins between villages and forests, are considered guardians of the mysterious forest realm, and widely believed to possess supernatural capabilities (Djabarudi 1990; Wessing 1995). As such, "the tiger in Southeast Asia is seen as an embodiment of the shaman spirit or soul and the vehicle *par excellence* of shamans in their quest for cures and access to the mysterious powers of the forest" (Wessing 1993: 6). Additionally, as ancestral human spirits and symbols of ruling authority (be it by the kingdom or court), tigers also possess the ability to maintain order in the human communities. According to Wessing (1995), tigers function to both preserve the ecological and spiritual order of the forest, and to guard human morality and *adat*, the "local traditional systems of rights, beliefs and customs" that governs life in the village (Henley & Davidson 2007: 3). With the tiger now extinct (for all intents and purposes), its magic properties and symbolic representation of nature's power has been transferred to the Javan leopard (*Panthera pardus melas*) – itself a critically endangered sub-species (IUCN Red List; Ario et al. 2008; Stein et al. 2020).[8]

With this brief introduction to the richness of the symbolic–historical–ecological landscape, it should now be clear why the natureculture lens, situated within an ethnoprimatological framework, is a befitting approach towards the study of the human–alloprimate interface in Java. Animals exist in the world, but the human–animal relationship is co-produced (and re-produced) through past, present and future interactions. In West Java, tigers (and now leopards) are perceived as symbolic ancestors that guard both the welfare of their descendants and the properties of the natural realms they inhabit. Furthermore, banteng (forest cattle) are associated with the establishment and welfare of states upon a healthy foundation of a wet–rice agricultural base (Wessing 1986, 2006) (Figure 4.2). Although these various species are either diminishing or locally extinct, their power comes from their "presence", of them being "there" – or at least a perception that they exist somewhere (Wessing 1994). These nonhuman animals are more than endangered nature, they are said to embody the spirits of their ancestors, and to rule over and protect the forest – overseeing order within and beyond the forest boundary. Without them, as one informant told us "the forest is destroyed … the sacred value is defiled" (Malone et al. 2014a: 45).[9]

Figure 4.2 Banteng (*Bos javanicus*) at the Cidaon grazing area, Ujung Kulon National Park. Photo credit: Nicholas Malone.

What about Java's primate species – do they feature in local mythologies? What does their presence (or absence) mean for the health of forests and that of the adjacent human communities? Over the past 20 years, my ethnoprimatological research engagements have sought answers to these questions. The extant community of nonhuman primates on Java includes: the silvery gibbon (*Hylobates moloch*); the grizzled leaf monkey or Javan surili (*Presbytis comata*); the Javan lutung (*Trachypithecus auratus*); the Javan slow loris (*Nycticebus javanicus*); and the long-tailed macaque (*Macaca fascicularis fascicularis*). With the exception of the macaque, each of these species has a very limited geographic distribution, and a current IUCN Red List status of either Critically Endangered, Endangered or Vulnerable.[10] In the next section, I will describe my research from two localities: in the Sancang Forest Nature Reserve in Propinsi Jawa Barat (West Java Province) and in Ujung Kulon National Park, Propinsi Banten (Banten Province).[11]

Leuweung Sancang

Upon the mention of having spent multiple nights (indeed months) in the Sancang forest, my conversations with Sundanese people, in villages and in cities, inevitably turn towards the topic of mysticism. Commonly, people ask me if I have experienced any supernatural phenomena. Did I use spiritual amulets (*jimat*) to protect myself, or intermediaries (*kuncen*) to unlock spiritual powers (*ilmu*, in this context)? And always – *are you familiar with the legend of Prabu Siliwangi?* As discussed above, forests play an important role in Sundanese mythology; however, the Sancang forest occupies an especially prominent location on the natureculture landscape. Legend has it, Prabu Siliwangi, the great ruler of the Sundanese Kingdom of Pajajaran – the pre-eminent Hindu state in West Java from 1535 to 1579, fled to the Sancang forest in order to avoid a confrontation with a close relative (Kean Santang) over the conversion from Hindu to Islam.[12] Wessing (1993: 16) writes:

> it is in this forest, the home of shamanistic powers, that Prabu Siliwangi changes into a tiger, the embodiment of his shaman spirit, and it is in this form that he is said to still watch over his descendants and the Sundanese generally.

A particularly descriptive attestation of the myth's ongoing salience was given to me by an experienced conservation official that had spent several years stationed at the guard house on the edge of the Sancang forest. He described to me, and then doctoral student Megan Selby, the following experience:

> So, what I remember that when we patrol at night we became lost ... [*indicates a pause*] We were circling in the middle of the forest until 10 o'clock at night. After we got tired, then came a tiger ... but it wasn't a real animal ... it is maybe a supernatural creature ... maybe it was somebody's

ancestor. After it appeared, suddenly the path was also seen by us. We were guided directly and escaped from our circling. That was about 11 o'clock at night ... that is what I remember.

The style and praxis of Islam that landed in West Java was strongly mystical, and, as such, fit well with the pre-existing substrate of beliefs (Wessing 1993). In deference to Prabu Siliwangi, generations of Sundanese people have made pilgrimages to Sancang to perform ritualistic practices that combine mysticism and Islam. These pilgrims seek to attain the promise of prosperity in difficult times. Spiritual visits to sacred sites (*keramat*) vary in length from several hours to several months or longer, and have traditionally been supervised by *kuncen* (Malone 2007; Reisland 2013; Malone et al. 2014a) (Figure 4.3). One pilgrim, a man in his early fifties from Jakarta, told me:

> According to my ancestor, from my grandfather until me ... I follow the ancestral custom directly upon entering the Sancang forest. They said the forest is special. Provided we act accordingly we will receive a good result.

Over the past decade, in correlation with a general trend of economic instability, there has been an increase in the number of pilgrim visits to the forest. However, from a deeper historical perspective, village elders describe the number of pilgrims as declining when compared to past generations. Pilgrim visits also ebb and flow according to particularly important nights on the Islamic calendar.[13]

Figure 4.3 Sacred sites, or *keramat*, within the Leuweung Sancang. On the left is the Cikajayaan waterfall, easily the most popular pilgrimage location within the forest. Pilgrimage visits vary in length from several hours to several months, with stated goals ranging in scope from simply bathing in the sacred waters, to attempts to "tilem – menghilang dari dunia" (disappear from earth, or become enmeshed with the spiritual world). On the right is an adult male silvery gibbon, as photographed from a less-frequently visited sacred site.

Photo credits: Nicholas Malone and Ajat Surtaja.

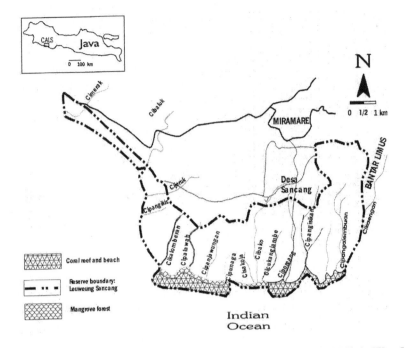

Figure 4.4 The location of the Cagar Alam Leuweung Sancang (CALS) in West Java, and a detail of the major geographical features within the reserve boundary (Malone & Wedana 2019).

In 1978, over 2000 hectares of the Sancang forest (*leuweung Sancang*) were set aside by the government to establish a nature reserve, or *cagar alam* – hence the local name *Cagar Alam Leuweung Sancang* (CALS) (Figure 4.4).[14] The strict conservation status served to protect this increasingly rare, lowland dipterocarp (family Dipterocarpaceae) forest, as well as a unique assemblage of endemic fauna. The site has now been the subject of multiple investigations primarily focused on the behaviour, ecology and conservation of the small population of silvery gibbons contained therein (Megantara 1995; Malone 2007; Reisland 2013; Malone et al. 2014a; Reisland & Lambert 2016). With extensive logistical support from collaborators and field assistants, primarily based in West Java's provincial capital of Bandung, I have studied the gibbon population at CALS during multiple fieldwork seasons between 2005 and 2016. In this research, I have used a combination of ethological observations (>300 hours of behavioural scan sampling and fixed-point counts of gibbon vocalisations), ecological monitoring (longitudinal assessment of habitat alteration over an 11-year period) and ethnographic methods (participant observation, cultural mapping and semi-structured interviews with more than 50 participants).[15] The duration of research visits to the site varied between one and seven months. Although longer field seasons typically yield data at a higher resolution, consistently returning to an anthropological field site permits a chronological account of salient changes, in addition to strengthening an array of important interpersonal relationships (Ramos 1990).

Despite both the official protections and the cultural significance, the forest has been dramatically altered by logging, forest conversion and other human modifications that greatly facilitate increased access and exploitation (e.g., motorbike paths, the construction of semi-permanent shelters, etc.). To date, approximately half of the forest within the reserve boundary has been lost. The majority of these losses occurred between 1999 and 2001, and are attributable to a period of large-scale timber extraction. Additionally, the ongoing small-scale, illegal extraction of valuable trees within the reserve boundary is evident (Malone & Wedana 2019). While walking through the forest one night in 2015, the son of a prominent kuncen told me:

> But long before when it's still Soeharto era … here the banteng still exist. The forest can be said *still haunted* [spoken with emphasis]. Before, the trees were big … all of the area from over there … dark … similar of the trees down in Cipangisikan [location with highest density of sacred sites]. But the trees close to the sacred site weren't cut down … no … forbidden … it was prohibited by the kuncen.

He continued to tell me that many spiritual pilgrims came during these times, and being a kuncen was a full-time occupation. However, intensive logging around the turn of the century made the eastern portions of CALS dramatically more penetrable, and there was a visible shift in the patterning of human behaviour with respect to the forest. One resident of the adjacent *Desa Sancang* (Sancang Village) succinctly captured the atmosphere, then and now: "in the past, it can be said that the forest had a haunted, ghostlike status … in the past … before there was the forest encroachment. Now we don't have a scared feeling like that". By all accounts, the local residents did not profit significantly from the logging of the valuable timber. The newly felled forest did provide residents the opportunity to expand agricultural fields and establish access trails to the coast, where fishing and the collection of seaweed are common.

Along with tigers and wild cattle, silvery gibbons were repeatedly mentioned in interviews, especially in regard to their songs that were commonly heard throughout the local villages. Participants in the research discussed that children today no longer wake to the sounds of the gibbons. The village leader, who now lives over a kilometre from the forest edge (due to deforestation), worries that there will come a day when his "grandchildren will not be able to recognise the call of the gibbons". The memory of a diverse community of wildlife resonates with locals for several reasons: it signifies the health of the forest, and, by extension, the health and stability of the community. A further decline in species richness and abundance could lead to a weakening of the forest's cultural allure (Malone et al. 2014a).

In addition to the loss of habitat, the illegal trade in silvery gibbons is ongoing, and poses a serious risk to the viability of silvery gibbon populations throughout their geographic range. This include CALS, where in 2008 a total of four gibbons were captured. Nijman (2005) calculated a potential loss of up to 6.86% from the wild population, per year, as a direct result of the illegal trade. Displaced silvery gibbons are found in all stages of the illegal trade network, including private homes, animal markets and wildlife rescue centres.[16] At the Javan Primates Rehabilitation Centre (JPRC), between 2011 and 2016 more than 30 silvery gibbons have been confiscated from private ownership (Wedana Adi Putra & Jeffery 2016). Of the rescued silvery gibbons, it is likely that, based on an assessment of the age, sex, physical appearance and details of their procurement and possession (as provided by the owners), at least one or two of the four gibbons that disappeared from CALS in 2008 have been recovered from the illegal trade (Wedana Adi Putra, unpublished data). Understanding the dynamics of the illegal trade in wild silvery gibbons, and subsequently their rehabilitation and (potential) re-release, remains the greatest challenge towards facilitating a survival strategy for this endangered species. I'll be discussing these topics further in Chapter 5.

Although we've documented forest loss and the illegal hunting of animals within CALS, the small population of silvery gibbons continues to persist.[17] Moreover, in comparison to other forests in Java, and with other areas within CALS itself, the presence of sacred sites and culturally enforced protocols appears to have safeguarded the gibbon habitat. In CALS, we find substantially higher group density co-occurring with a higher density of sacred sites and pilgrim visits (Malone 2007; Malone & Wedana 2019; and Chapter 3 – *Hylobatid Community Dynamics*). Importantly, the forest and its denizens represent connections to salient religious and social facets of Sunda's cultural history. For the residents of Sancang village and the majority of the pilgrims, a forest filled with wildlife is perceived as a corporeal connection to their own human spirits, and thus the survival of animals is a link to the survival of their own ancestors. Indeed, for numerous residents, the presence of silvery gibbons signifies a forest imbued with life, health and spirituality (Malone et al. 2014a). Finally, during an interview with a particularly pious, 39-year-old man during his pilgrimage to Sancang, we discussed his experiences in Java's sacred forests. Describing himself as a *musafir* (traveller/wanderer), he sought not wealth or success, but instead desired the ability to become invincible – the "kind of magic that is directly handed down from a tiger". In his travels, he had previously visited both Alas Purwo and Ujung Kulon, expansive national parks on the easternmost and westernmost tips of Java Island, respectively. When asked for his assessment of these famous forests, relative to Sancang, he unequivocally referred to the "supreme power" of Ujung Kulon, due to its natural and spiritual "completeness". There he practised what he referred to as the "essence the musafir", that is:

> to control the carnal desire … especially the fear when we are in certain place … especially in the forest. Fear of hunger, fear of wild animals and

fear of haunted creatures. If other people feel afraid … on the contrary we feel strengthened. We learn to control the fear lest the fear conquer us. Because the fear is a part of carnal desire. Who are we more afraid of: God or the surrounding environment?

In 2019, after three years of planning and anticipation, I made my first trip to Ujung Kulon.

Ujung Kulon

At 9:30 PM local time, on the night of 22 December 2018, tsunami waves began emanating from the south-western flank of a volcano in the middle of the Sunda Strait. Approximately 14 minutes later, a series of waves began wreaking havoc upon the low-lying, coastal communities of southern Sumatra and westernmost Java (Syamsidik et al. 2020). More than 400 people died in the event, and nearly 10 times that amount were either injured or displaced (Figure 4.5). Under the radar of alert systems which are designed to pick-up seismically generated tsunamis, the eruption and structural collapse of the volcano Anak Krakatau brought destruction, in the form of tsunami waves up to 6 metres in height, upon an unsuspecting populace.[18] When I arrived six months

Figure 4.5 The effects of tsunami waves on the night of 22 December 2018 on both the peninsula and mainland portions of Ujung Kulon National Park. The photo on the left, near Citelang, reveals the extent of destruction to coastal trees (extending approximately 150 metres inland from the inter-tidal zone). The photo on the right is of government-provided housing, near Tamanjaya, for coastal-wary and/or displaced communities. With only short distances between the sea and the protected reserve boundary, a difficult balance is being struck between safety, livelihoods and ecosystem impacts.

Photo credits: Nicholas Malone.

later, the physical and psychological scars of the tsunami were readily apparent. One of my first interviews was with a park guard, a man in his late-fifties who was stationed at Citelang, a remote portion of Ujung Kulon, and one the hardest hit by the tsunami event. He spoke solemnly of his narrow escape:

> Less than half-an-hour after the eruption at Krakatau I heard what sounded like a helicopter landing. I could see the crest of the wave approaching [contorts his right wrist into the shape of a snake-like curl]. I climbed into a tree outside the guard house ... fortunately the tree had ladder-like steps. The 4-metre tall wave came in and wet my feet, but thanks be to God, I was given more life. Only four of us from the six [men stationed at the outpost] survived.

Anak Krakatau is of course the child of Krakatau[19], whose eruption in 1883 is considered to be not only one of the largest in recorded history, but whose effects were experienced, both climatically and consciously, by an increasingly connected global community (Winchester 2004). However, it is the local effects of both the 1883 and 2018 volcanogenic-tsunami sequences that concern us here; for only a short distance away (~60 kilometres) lies the most complete assemblage of Javan wildlife, ranging within the extensive lowland forests of the peninsula and adjacent mainland (Whitten et al. 1996). With a strategic location overlooking the Sunda Strait, and an abundant amount of game to hunt and "indiarubber" to export (latex, from wild specimens of *Ficus elasticus*), Ujung Kulon "seemed predestined to become one of the most important parts of Java" (Hoogerwerf 1970: 9).

In the lead up to the 1883 cataclysm, a low density of human settlements, supported by shifting cultivation, lined the coastal areas of this remote forest. As Java's population continued to rise, the exploitation of the peninsula's natural richness was slowly intensifying, as was game hunting (Whitten et al. 1996). In 1883, the obliteration of Krakatau, the significant accumulation of ash, and the associated tsunami waves intervened to re-calibrate the relationship between human livelihoods and ecosystem process. The event led to a dramatic reduction in human exploitation and a chance for wildlife populations to stabilise. At the time, the thriving animal community consisted of: herbivores (banteng and rhinoceros), all five of Java's primate species; predatory cats; and a further array of small mammals, birds and reptiles. However, the abundant wildlife attracted game hunters, and concerns over depletion led to calls for some level of protection. A portion of the Ujung Kulon peninsula was declared a nature reserve by the Dutch colonial government in 1921. Subsequent decrees and proposals between 1937 and 1980 resulted in the extension of protected boundaries and the amalgamation of adjacent reserves resulting in a more formalised conservation area that included: the peninsula; an array of offshore islands and designated marine areas; and the Gunung Honje ranges on the mainland. Most notably, in 1992 Ujung Kulon was granted National Park status by Decree 284/Kpts-II/ 1992, and in the same year, coupled with the Krakatau Nature Reserve, was placed on the UNESCO World Heritage List under Natural Criteria vii (aesthetic) and x (endangered species of universal value). The present size of Ujung Kulon National Park (UKNP) is 122,956 hectares (ha) comprising 78,619 ha of mainland and 44,337 ha of marine habitat (Figure 4.6).[20]

UKNP's status as a World Heritage site and the refuge of several notable, endangered species (including gibbons, rhinoceros and banteng), ensures its continued prominence, attracting the attention of international tourists and conservationists. However, with approximately 50,000 people residing within the park's buffer zone (±19,500 ha), the boundaries between nature and society are perpetually re-negotiated among the various stakeholders. In June 2019, I joined with colleagues from a variety of disciplines (sociocultural anthropology and biology) at the Universitas Padjajaran, a leading Indonesian university based in Bandung, West Java. Our goal was to establish a research programme within

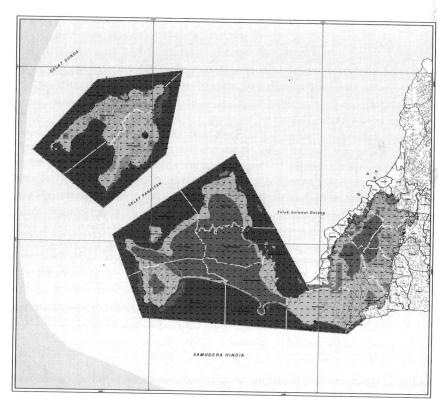

Figure 4.6 The present boundaries and major regions of Ujung Kulon National Park, including (from left to right): Panaitan Island, the peninsula and the mainland (including the Honje ranges). The most prominent areas on the map indicate the Jungle (*light grey*), Core (*medium grey*) and Marine Protected (*dark grey*) Zones.

Map courtesy of Balai Taman Nasional Ujung Kulon, Labuan, Banten.

the UKNP and the bordering communities. Over a two-month period, we conducted a preliminary survey of silvery gibbon habitat using repeated, line-transect sampling to assess gibbon population parameters on the western slopes of Mt. Honje. Additionally, 12.5 kilometres of reconnaissance walks were used to estimate the extent of recent, small-scale logging of semi-evergreen rainforest within the park boundary. Finally, ethnographic insights were obtained by participant observation and semi-structured interviews with a variety of stakeholders including rice farmers, local officials and spiritual pilgrims.

During June and July 2019, our research team established four, 2 kilometre-long transects on the western slopes of the Gunung Honje ranges (specifically in the Tamanjaya management section). A fifth transect was established to the south in the Legon Pakis management section, in order to ensure sampling of the Cimahi pilgrimage site. Each of the five transects were sampled three times over the course of six weeks by walking at a pace of approximately 0.5 km/hour, in a west to east direction between 5:30 and 9:30 AM. A sighting distance of 50 metres on either side of the transect line was deemed reliable and used in the calculation of density estimates. Overall, we estimated an average gibbon group density at 3.65 groups/km^2 and an average group size of 2.55 individuals/group. Therefore, we calculated population density to be 9.30 individuals/km^2. Previous research in the Honje ranges section of the park provides some comparable insight. Iskandar (2001), using similar methods to survey the Gunung Honje – Cibiuk area, reports a lower group density of 2.80 groups/km^2, but a higher, average number of individuals per group (2.80). Therefore, Iskandar's (2001) calculation of individual density (9.20 individuals/km^2) is virtually identical to our estimates (Malone et al. 2020). Some of the gibbons we observed are not averse to ranging in localities that include concentrated forms of human activity. Specifically, the density of gibbons was highest in our transect that included the sacred (*keramat*) location of Cimahi, a popular pilgrimage site within Legon Pakis. In parsing the data to compare densities in pilgrimage versus non-pilgrimage locales, we uncovered a pattern that resembles the distribution and densities of gibbons in the sacred Sancang forest (CALS) (Table 4.1).

In both structural and aesthetic ways, Cimahi in Ujung Kulon is remarkably similar to the Cipangisikan pilgrimage area in CALS (Figure 4.7). Like Sancang, visitors to Cimahi connect with spiritual guides (*kuncen*) and arrange for any required logistics in the adjacent village. After a moderately strenuous 60–90-minute walk, pilgrims arrive at a particularly cathedral-like area of primary-growth forest. Upon arrival, pilgrims settle-in to one of several small shelters where, for variable periods of time, prayers and meditation are interspersed with conversation, bathing and victuals. Local people recognise Cimahi as a *makam keramat* (sacred burial ground) and one of the healthiest examples of animal habitat in Ujung Kulon National Park. In the past, before many people began to make Cimahi a location of frequent pilgrimage, the local people of Ujungjaya Village believed that if either injured wild animals or sick people visited this place, they would recover because of the magical forces in this area

Table 4.1 Comparison of population parameters for silvery gibbons in the Cagar Alam Leuweung Sancang (CALS) and Ujung Kulon National Park (UKNP). At both sites, higher group and individual densities are associated with the common presence of spiritual pilgrims

Area	Location/ Description	Avg. Group Size (Indivs/group)	Density (Groups/km²)	Density (Indivs/km²)	Citations[a]
CALS	Cipangisikan/ Pilgrims common	3.30	3.00	9.90	1, 2
	Cipalawah/ Pilgrims rare	3.30	0.50	1.65	1, 2
UKNP	Legon Pakis– Cimahi/ Pilgrims common	3.00	4.15	12.45	3
	Honje– Tamanjaya/ Pilgrims absent	2.33	2.49	5.80	3

a 1. Malone (2007); 2. Malone and Fuentes (2009); and 3. The present study.

Figure 4.7 The pilgrimage site of Cimahi in the Legon Pakis management area of Ujung Kulon National Park. Pictured at right is the grave of a saint (one of five) in this *makam keramat* (sacred burial ground).

Photo credits: Nicholas Malone.

(Permana et al. 2020). Today at *makam keramat* Cimahi there are five graves of saints, namely: *Uyut* or *Syaikh Santika*, *Syaikh Dahlan*, *Uyut Nuralam*, *Uyut Marga* and *Uyut Bidadari*. Local people and pilgrims are prohibited to disturb or hunt primates, who are considered as guardians of the graves. Failure to follow this prohibition would invite disaster and prevent their wishes from being granted.

Pilgrims visiting *makam keramat* Cimahi are aware of the primates' presence, and often observe the animals moving and feeding in the canopy above.

Another component of our research was to identify if any "primate-themes" were embedded in local myths, stories and traditions. In our research, we confirmed a common belief among people residing in villages surrounding Ujung Kulon, that the silvery gibbon (*kuweung* in the local dialect) originated when a pair of orphaned children were banished to the forest (*leuweung*) by their uncle. An elderly participant, a local farmer, provided the complete story in detail.[21] Because silvery gibbons originated from humans, they were not to be disturbed or harmed (Permana et al. 2020). Myths and beliefs underpin a variety of behavioural interactions between humans and wild animals, including conflicts, taboos and forms of utilisation (e.g., Nijman & Nekaris 2015). In related research, rural Sundanese communities who inhabit the Cisokan watershed in West Java possess a shapeshifting myth where endemic primate species, including the silvery gibbon and the Javan surili (*Presbytis comata*) can morph into a Javan leopard (*Panthera pardus melas*). Because of this, certain attitudes commonly associated with leopards were extended to the primates (Permana et al. 2019). However, in a more general sense, the conservation efficacy of indigenous knowledge is vigorously debated (see Maffi 2004).

The population of silvery gibbons within UKNP is one of three remaining population strongholds for the species. In a population viability analysis, Jaima Smith and colleagues calculated the extinction probability over a 100-year time period under various scenarios involving variable degrees of deforestation and loss of gibbons due to illegal hunting/trade. Smith et al. (2017) determined that the gibbon population at UKNP is indeed viable, provided their estimates for annual hunting losses can be avoided at all costs. Unfortunately, as I have recorded at Sancang, relatively widespread reverence for the forest and its denizens does not produce an infallible immunity from illegal poaching. As such, the prevention of losses from hunting to supply the illegal pet trade remains the top conservation priority. With respect to deforestation, a scenario involving modest losses (1.2% annually) will not result in extinction, but lead to a reduced, remnant population ($n = 135$) within the 100-year time frame of the model (Smith et al. 2017). Although the results from our 12.5 kilometres of reconnaissance walk "ground-truthing" exercise remain preliminary (Malone et al. 2020), the observation of newly created irrigated rice paddies and ongoing small-scale deforestation within the park boundary are deeply concerning (Figure 4.8). Return trips to compare against this baseline, I suspect, will confirm the present pace of forest loss will either match or exceed the modelling estimates. In sum, our ethological and ethnographic research findings provide the basis for a mix of both concern and cautious optimism in relation to the endangered wildlife within UKNP. Can local traditions, including historical connections to sacred forests and the animals within, be empowered by ensuring continued access to the forest? Or are the pressures of an expanding populace too great to prevent continued deforestation beyond the threshold identified in viability analyses?

Figure 4.8 Reconnaissance walks along the forest-field transitional boundaries to map the extent of agricultural encroachment into the National Park.
Photo credits: Nicholas Malone.

It is clear that hunting losses need to be prevented, yet questions remain about how this goal can be accomplished.

So where to from here?

To summarise, both CALS and UKNP harbour small, but potentially viable, populations of an endangered and endemic ape species. Sacred sites nestled within both forests have consistently attracted spiritual pilgrims from throughout West Java. Such mysticism gives these forests a particular allure. However, despite the official protections and cultural significance, the forests have been dramatically impacted by human actions that facilitate access and permit exploitation. The populations of silvery gibbons (and other, significant mammalian taxa) continue to persist, but illegal hunting and habitat degradation may ultimately lead to their demise. While my previous research has emphasised the ecological and demographic aspects of the focal species' space, the present research trajectories at the sites explore the spiritual elements of these shared, multispecies places. In these investigations I endeavour to more fully understand local socioeconomic structures and belief systems. Cultural beliefs and livelihood practices are shaping factors of landscapes in ways that have vital implications for the resilience of ecological and social systems. In short, these landscapes are as much socially constructed as they are material (Malone 2007; Malone et al. 2014a, 2014b). As discussed in this chapter, silvery gibbons demonstrate a response sensitivity to human activities, and therefore effective conservation strategies for this endangered species depends upon our continued understanding of behavioural ecology within human-modified habitats (Reisland & Lambert 2016; Malone & Wedana 2019). Certainly, the situation on Java is distinct, but it is not unique. I contend that with a parallel research lens, similar natureculture dynamics would be identified as salient for many, if not all, extant primate species.

At Ujung Kulon, tigers (through memories of their existence), gibbons, banteng and rhinoceros are symbols of a healthy forest; and in turn a healthy forest is indicative of well-balanced societal relations. At Sancang, the site of Prabu Siliwangi's transformation into a spiritually powerful form, the forest becomes a physical-symbolic border between nature and society, as well as between historical eras (Wessing 1993, 2006). Social scientists and historians occasionally focus on the relationships between humans and other primates (Janson 1952; Corby & Theunissen 1995), and the analysis of other animals' roles in human symbol and myth has been a facet of social anthropology since the 1950s (Lévi-Strauss 1963; Shanklin 1985; Ingold 1988). O'Flaherty (1988: 35) argues: "myth is not so much a true story as a story on which truth is based, a story which people may infuse with their truth". The continued functioning of the forests in these ways sustains both economic livelihoods and ecological relationships. In Java, the vitality of both forests and of local communities depends upon the (symbolic and material) persistence of biodiversity. By all accounts, the historical and contemporary engagements among human and primate ecologies are largely inseparable. As such, models of research and conservation must account for the entangled nature of the relationships among species – be they material or imagined.

While the efficacy of myths and indigenous knowledge – in relation to the management of natural resources and the protection of biodiversity – will continue to be debated among anthropologists and conservationists, I hope to have made a more general point abundantly clear: enhanced elucidation of the interconnectivity of natural and social systems can only be achieved through non-reductionist means. Adger et al. (2005: 1036) define conservation resilience as "the capacity of linked social-ecological systems to absorb recurrent disturbances … so as to retain essential structures, processes, and feedbacks". Given the complexity of issues surrounding both the CALS and the UKNP-protected areas, I have argued that shifting from a targeted "species conservation" paradigm to a "resilience" framework establishes a realistic lens through which to envision a future for the forest and its denizens (Malone & Wedana 2019). Going forward, facilitation of the ape–human interface ultimately requires the establishment of strong relationships with people who are dependent on forest resources for their livelihoods and subsistence needs (e.g., food, medicine and materials for housing construction). Identifying the interests of local communities (which are, in all likelihood, not identical to those of ape conservationists) begins the process, but further work is required to move diverse agendas towards compatible goals. Chua et al. (2020: 53) recommend the use of "proxies" (i.e., "issues that can stand for different parties' concerns and constitute a shared point of engagement between them") to create space for productive collaborations. It is unlikely, and potentially counterproductive, to seek a ban on all human activities within most primate habitats. Instead, striving to identify synergies may guide local management decisions, and facilitate resiliency within these linked, ape–human social-ecological systems.

It should be clearly noted that identifying salient natureculture dynamics *is different* from a generic promotion of so-called integrated conservation and development projects. In his 1999 book *Myth and Reality in the Rain Forest*, John Oates argues convincingly against the increasingly promoted position that efficacious wildlife conservation is achieved by linking biodiversity conservation with the economic development of local people, or "communities". In both theory and practice, Oates' position contrasts with those that see economic development as "not inherently antithetical to the conservation of nature and that rural people will tend to be more protective of wildlife if they are given greater control over it" (Oates 1999: xii). Instead, Oates identifies several key issues with granting of greater control of wildlife management to rural peoples, including the assumptions that rural people: a) form cooperative communities where self-interests are minimised; b) use natural resources in more sustainable ways; and c) maintain long-term ties to the land they presently occupy. Based on experiences, primarily in the rain forests of West Africa, Oates encourages conservation organisations to return to setting agendas based on the intrinsic value (ethical and aesthetic) of wildlife as opposed to the consideration of human economic well-being. While I am sympathetic, in part, to Oates' argument, I see the importance for the people of Java, and likely elsewhere, to retain material and spiritual connectedness to the aesthetic aspects of forests, and the species of wildlife therein. In Java, the well-being of communities and the regulation of forest resource utilisation, as facilitated by mediated access, are not mutually exclusive.

And herein lies the core of the dialectical analysis as applied to primate conservation. Whereas the principles of emergence, constrained totipotentiality and constructed compatibility (or adaptive complementarity) have been shown to be of evolutionary import (see Chapter 3), these same concepts are also deployable within the realm of conservation management. Similar to how once vaunted dichotomies (e.g., Mayr's proximate-ultimate distinction) can divide labour and limit theoretical developments (Laland et al. 2011), framing conservation issues in oppositional terms ignores relational exchanges and reciprocal processes. In the anthropocene, we increasingly lack the ability to separate the evolutionary trajectories of endangered species from humankind's rapid transformation of entire ecological systems. The stabilisation of critical relationships requires us to identify an emergent synthesis from the truths and errors of oppositional assertions, a hallmark of a dialectical analysis. Somewhat paradoxically, the search for synthesis is more about the questions than the answers. As demonstrated in this chapter, my experiences in Java have undoubtedly forged the "habits of thought" I carry forth as a reflexive, anthropological primatologist. The full extent of cultural, political and historical complexity weighs upon one's research and/or conservation activities. These inherent, internal dialectics structure the data and insights which are obtained. Attending explicitly to these dialectics shapes the way future projects are conceived: the way future questions are framed. Whether one refers to this approach as "ethnoprimatology" or the "natureculture" framework (or indeed a dialectical primatology) is somewhat arbitrary; the tie that binds is the attention to connected entities, agencies and contradictions within a complex system. These insights, in conjunction with more traditional primatological assessments of what influences ape behaviour, provide us with a more complete understanding of the critical elements within human–ape interfaces. As we'll see in the next chapter, a full awareness of the constructed nature of our relationships with apes, be they wild, captive or somewhere in between, will be utterly consequential for the future of hominoid life.

Notes

1 What I am glossing as "patience" is really the ability to regulate one's internal and external states, and I have observed its practice in my informants, field assistants, acquaintances and village hosts. I have attempted to emulate this ability, as Geertz (1975) described, to "flatten out … the hills and valleys of emotion into an even, level plain" (49).

2 This species/area relationship persists statistically, albeit less robustly, with the inclusion of the Mentawai Islands and their associated high numbers of species and genera (and endemism) (Harcourt 1999).

3 In Jablonski (1998), a spatial database was used to demonstrate distributional changes for each of the recognised sub-epochs of the Pleistocene and the Holocene: Early Pleistocene (2.5 mya–780 kya), Middle Pleistocene (780–128 kya), Late Pleistocene (128–11 kya) and Holocene (11 kya–present). A second database recorded the occurrence of target genera in specific locations at specific times. The distribution of the fossil occurrences of five genera (*Gigantopithecus, Pongo, Hylobates, Macaca* and *Rhinopithecus*) was plotted against the paleo-environmental conditions in China during the aforementioned time periods.

4 "… the last one having been shot, it is said, by either President Suharto of Indonesia, Prince Bernhard of the Netherlands, or the late Shah of Iran" (Wessing 1995: 192).

5 As previously mentioned, the Javan tiger (*Panthera tigris sondaica*) is extinct. The remaining rhinos on Java (*Rhinoceros sondaicus sondaicus*) number 68 (personal communication with head of Balai Taman National Ujung Kulon in Labuan, Pandeglang, Banten on July 2019), and the Endangered wild bull, or banteng (*Bos javanicus*) is limited to four or five sub-population "strongholds" (w/ > 50 individuals each) (Gardner et al. 2016).

6 The utility of the natureculture concept is evidenced by a brief example from the adjacent island of Bali. Balinese people and long-tailed macaques (*Macaca fascicularis*) have lived in sympatry for over a thousand years, and in this time have together developed a mutual ecology where the lives of Balinese people and macaques are deeply inter-twined (Wheatley 1999; Fuentes 2010). The more recent surge of tourism in Bali provides additional layers of relationships. The Balinese people–macaque–tourist interface, viewed through the naturalcultural framework, incorporates dietary, economic, parasitological, religious, behavioural, political and geographical elements (Fuentes 2010). In Malone and Ovenden (2017: 1–2), we used the natureculture lens to outline the human–alloprimate interface in Bali as follows: (1) macaque populations that congregate at Hindu temple sites, where they are ascribed a sacred, protected status, which partially underpins population growth; (2) a tourism industry that facilitates the existence of the politically and

economically driven tourist–macaque interface; (3) the livelihoods of Balinese people who are financially supported by tourist–macaque interactions; (4) macaque diets that are being modified by tourist provisioning; and (5) belief systems that influence the mutual ecologies of the various actants at these sites. Integrating these elements enables holistic investigations of co-constructed, multispecies relationships.

7 Sometimes boundaries can be both material and conceptual. Occupying such a conceptual/material hybrid–boundary are animals on the brink of extinction: they are confined to the remaining margins of suitable habitat; and they are subjected to debates concerning their value and/or relevancies, at both the individual and taxon level (Lowe 2004; Malone et al. 2014a).

8 As described in Wessing (1995), some linguistic ambiguity facilitates this transference of symbolic power from tigers to leopards, as both are frequently referred to using the same word (*macan*) by peoples across Java. Additionally, several iterations exist that incorporate modifiers to reflect both morphological variants and folk taxonomies.

9 I should note that not all engagements with forested habitats are characterised as "sacred" by our informants. In other contexts, a more utilitarian view predominates, such as how farmers describe the value of forested watersheds (see Chapter 5).

10 Silvery gibbon, Endangered (Andayani et al. 2008); Javan surili, Endangered (Nijman & Richardson 2008); Javan lutung, Vulnerable (Nijman & Supriatna 2008); Javan slow loris, Critically Endangered (Nekaris et al. 2013); and the long-tailed macaque, Least Concern (Ong & Richardson 2008).

11 Founded two days after Indonesia's national independence in 1945, West Java (Propinsi Jawa Barat) originally extended from Central Java in the east to the westernmost extent of the island. The capital city of Jakarta was partitioned out as a special administrative region in 1966. In 2000, the westernmost portion of Java Island, known as Banten, was separated from Jawa Barat to establish an independent province. While generally sharing legislative, economic and cultural histories, particularities abound in the form of linguistic and religious nuance among these respective regions. My research in West Java was conducted with permission from the Indonesian Institute of Sciences (Lembaga Ilmu Pengetahuan Indonesia) and the Ministry of Research and Technology (Kementerian Riset dan Teknologi, or RISTEK), permit numbers: 4241/SU.3/KS/2003; 4000/SU/KS/2005; 1041/FRP/SM/VIII/2012; 921/FRP/SM/IIi/2013 and 238/SIP/FRP/SM/VIII/2015. Most recently, research in Banten Province received the following ethical clearances, visa and research approvals: University of Auckland Human Participants Ethics Committee (013475); RISTEK 306/FRP/E5/Dit.KI/III/2019; 190/E5/E5.4/SIP/2019 and a SIMAKSI from BKSDA: S524/T.12-TU/P3/06-07/2019.

12 There is variation in the various accounts as to the nature of the close relationship – some versions of the myth describe the conflict as being between a father (Prabu Siliwangi) and his son (Kean Santang), while other documents indicate that Kean Santang was Prabu Siliwangi's older brother and chief minister (see Wesssing 1993: 2, Note 13). Wessing interprets Kean Santang to be an idealised version of Sunan Gunung Jati, who founded the Sultanate of Banten (and indeed converted much of West Java to Islam). Both figures, it is said, travelled to Mecca and returned to convert West Java to Islam. The fact that shamanistic practices imbue the Islam of West Java aligns well with the notion that both Kean Santang and Sunan Gunung Jati possessed great magical powers (Wessing 1993).

13 Briefly, I'll mention two examples. The first is *Rajab* – the seventh month on the Islamic calendar and a time of loyalty and prohibition (Damanhuri 2016). The

second is *malam Jumat Kliwon* – an auspicious night according to Islamic oral tradition where supernatural forces are at their most powerful.

14 Over 2157 hectares of the Sancang forest located on the south coast of West Java
near Pameungpeuk in Garut Regency (S 07°43′30″, E 107°52′30″) was established
in 1978 (Decree: 370/KPTS/UM/6/1978).

15 Cultural mapping involves the collection of social, historical and ecological data
in situ, and facilitates a researcher's understanding of the importance of cultural
landscapes as active components in people's lives (Bender 1993; Morphy & Morphy
2006). The movement through places is one of the most important elements of cultural mapping as it often adds more detail and sometimes more information than
what can be gained during a stationary interview (Strang 2010).

16 Protection of primates in Indonesia began with the prohibition of hunting and
killing certain species by ordinance in 1925, when the country was governed by the
Dutch colonial administration. Included in this limited list were all known species
of gibbon and the orangutan (identified as *Simia satyrus*). Article 21 (Clause 2.a) of
the Act of the Republic of Indonesia No. 5 (1990) Concerning the Conservation
of Natural Resources and their Ecosystems – known widely as Undang-Undang
(UU) Nomor 5 Tahun 1990 – states that:

> *"Setiap orang dilarang untuk: menangkap, melukai, membunuh, menyimpan, memiliki,
> memelihara, mengangkut, dan memperniagakan satwa yang dilindungi dalam keadaan
> hidup."* Translation: Every person is prohibited from capturing, injuring, killing,
> storing, possessing, maintaining, transporting and trading protected animals in
> a live state.

Fines of up to IDR 100.000.000 (~$6750 USD), and prison sentences of up to five
years, can be imposed on lawbreakers. See also Chapter 5, Note 7.

17 Although a number of recent births have mediated a history of disappearances and
deaths (attributed to poaching or other unknown causes), there has been a modest
reduction over the past decade in the effective (but not overall) population size.
Consideration of the effective population size (N_e) is important as it is an indication
of the size of the breeding population, and therefore is more sensitive to the smallest
population sizes over time. Indeed, even if the census size of the population (N) can
be increased, the genetic variation may continue to decrease, because N_e still reflects
any recent bottlenecks. (Malone et al. 2014a; Malone & Wedana 2019).

18 The 2018 eruption triggered tsunami waves that impacted both the southern coast
of Sumatra (Lampung Province) and the western coast of Java (Banten Province).
Although the volcano is located approximately equidistant to these coasts, the wave
height and intensity were higher in Banten and resulted in more severe damage.
Because the tsunami event occurred at night, and the fact that it was not set in
motion by an earthquake, the inadequacies of the present, tectonic-driven warning
system were laid bare (Syamsidik et al. 2020).

19 Literally and figuratively, the child of Krakatau (in translation) and in the geophysical forces which have produced the emerging lava dome where the parent
volcano once stood. By 1927–1928, reports began to document the emergence of
the rebuilding volcano in the collapsed crater of its predecessor, eventually gaining
permanent visibility above sea level by 1930. Anak Krakatau continues its active
building and eruption phases laying testament to the unrelenting tectonic forces of

the Circum-Pacific Belt, commonly known as the Ring of Fire, and its associated subduction zones.

20 The 2,500 hectare Krakatau Islands Nature Reserve is managed separately by BKSDA, Lampung, Sumatra.

21 The full story was related to us as such:

> Once upon a time, a brother and sister (orphans) were left to die by their parents. But before their parents died, the father commanded of his older brother to take care of his children and property, and to give the property to the children when they grew older — *"ka ieu anak kula titip sabab kula gering bisi kula teu umurna, ieu anak kula urus jeung pangaboga kula rawat ku kaka, engké anak kula gedé bikeun"*. The older brother was a rapacious and cunning character, and wanted to dominate the property that was left for his younger brother. Instead of caring for the children, he banished his nephew and niece to the forest. He built a treehouse (*ranggon*) in the forest. The uncle commanded his nephew and niece to go up into the treehouse, and not to come down to the ground. The uncle set about cooking rice (*liwet*) intended for the nephew and niece in a bamboo grove. Secretly, the uncle left the forest. After three months had passed, the nephew and niece called their uncle, "*uwa* come back to *saung* (small house) because the rice has been burnt" (in Sundanese language, *uwa* is a term for the older brother of a father or mother). After that, the children's bodies were transformed: from pores of skin grew long hair, and they ate anything [including] fruit growing in the trees.

> (Permana et al. 2020: 523)

References

Adger WN, Hughes TP, Folke C, Carpenter SR, Rockström J. 2005. Social-ecological resilience to coastal disasters. *Science*, 309(5737): 1036–1039.

Andayani N, Brockelman W, Geissmann T, Nijman V, Supriatna J. 2008. *Hylobates moloch*. *The IUCN Red List of Threatened Species* 2008: e.T10550A3199941.

Ario A, Sunarto S, Sanderson J. 2008. *Panthera pardus* ssp. *melas*. *The IUCN Red List of Threatened Species* 2008: e.T15962A5334342.

Applegate G, Smith R, Fox JJ, Mitchell A, Packham D, Tapper N, Baines G. 2002. Forest fires in Indonesia: Impacts and solutions. In Colfer CJP, Resosudarmo IA (Eds.), *Which Way Forward: People, Forests and Policymaking in Indonesia* (pp. 293–308). Washington, DC: Resources for the Future Press.

Baldwin JR. 2003. *Reconceiving Wealth for Geographic Analysis*. PhD Dissertation, University of Oregon, Eugene, OR.

Balmford A. 1996. Extinction filters and current resilience: The significance of past selection pressures for conservation biology. *Trends in Ecology & Evolution* 11(5): 193–196.

Bender B. 1993. *Landscape, Politics and Perspectives*. Oxford: Berg.

Boomgaard P. 1988. *Statistik Indonesia 1987*. Jakarta: Biro Pusat Statistik.

Brooks TM, Pimm SL, Collar NJ. 1997. Deforestation predicts the number of threatened birds in insular Southeast Asia. *Conservation Biology* 11(2): 382–394.

Chua L, Harrison ME, Fair H, Milne S, Palmer A, Rubis J, Thung P, Wich S, Büscher B, Cheyne SM, Puri RK. 2020. Conservation and the social sciences: Beyond critique and co-optation. A case study from orangutan conservation. *People and Nature* 2(1): 42–60.

Colfer CJP, Resosudarmo IA. 2002. Introduction. In Colfer CJP, Resosudarmo IA. (Eds.), *Which Way Forward: People, Forests and Policymaking in Indonesia* (pp. 1–19). Washington, DC: Resources for the Future Press.

Corbey R, Theunissen B. 1995. Ape, man, apeman: Changing views since 1600. Evaluative proceedings of the symposium Ape, man, apeman: Changing views since 1600, Leiden, the Netherlands, June 28–July 1, 1993. Leiden: Department of Prehistory, Leiden University.

Dake A. 2006. *The Sukarno File, 1965–1967: Chronology of a Defeat.* Boston, MA: Brill.

Damanhuri D. 2016. The hadith of Rajab fasting. *Ar Raniry: International Journal of Islamic Studies* 3(1): 221–248.

Djabarudi S. 1990. Dari kiai untuk kiai [From venerated teachers to venerated teachers]. *Tempo* 20(24): 5.

Dore KM, Riley EP, Fuentes A. (Eds.). 2017. *Ethnoprimatology: A Practical Guide to Research at the Human-Nonhuman Primate Interface* (Cambridge Studies in Biological and Evolutionary Anthropology). Cambridge: Cambridge University Press.

Dudley RG. 2002. Dynamics of illegal logging in Indonesia. In Colfer CJP, Resosudarmo IA. (Eds.), *Which Way Forward: People, Forests and Policymaking in Indonesia* (pp. 358–380). Washington, DC: Resources for the Future Press.

Estrada A, Garber PA, Rylands AB, Roos C, Fernandez-Duque E, Di Fiore A, … & Rovero F. 2017. Impending extinction crisis of the world's primates: Why primates matter. *Science Advances* 3(1): e1600946.

Franke R. 1977. Miracle seeds and shattered dreams in Java. *Readings in Anthropology* 1: 197–201.

Fuentes A. 2010. Naturalcultural encounters in Bali: Monkeys, temples, tourists, and ethnoprimatology. *Cultural Anthropology* 25: 600–624.

Fuentes A. 2012. Ethnoprimatology and the anthropology of the human–primate interface. *Annual Review of Anthropology* 41: 101–117.

Fuentes A, Wolfe LD. (Eds.). 2002. *Primates Face to Face: The Conservation Implications of Human-Nonhuman Primate Interconnections.* Cambridge: Cambridge University Press.

Gardner P, Hedges S, Pudyatmoko S, Gray TNE, Timmins RJ. 2016. Bos javanicus. The *IUCN Red List of Threatened Species* 2016: e.T2888A46362970 (accessed on April 17, 2020).

Geertz C. 1960. *The Religion of Java.* Chicago, IL: University of Chicago Press.

Geertz C. 1963. *Agricultural Involution: The Process of Ecological Change in Indonesia.* Chicago, IL: University of Chicago Press.

Geertz C. 1975. On the nature of anthropological understanding: Not extraordinary empathy but readily observable symbolic forms enable the anthropologist to grasp the unarticulated concepts that inform the lives and cultures of other peoples. *American Scientist* 63(1): 47–53.

Haraway DJ. 2003. *The Companion Species Manifesto: Dogs, People, and Significant Otherness* (Vol. 1). Chicago, IL: Prickly Paradigm Press.

Harcourt AH. 1999. Biogeographic relationships of primates on South-East Asian islands. *Global Ecology and Biogeography* 8(1): 55–61.

Harcourt AH, Schwartz MW. 2001. Primate evolution: A biology of Holocene extinction and survival on the southeast Asian Sunda shelf islands. *American Journal of Physical Anthropology* 114: 4–17.

Harper J. 2002. *Endangered Species: Health. Illness and Death among Madagascar's People.* Durham: Carolina Academic Press.

Henley D, Davidson JS. (Eds.). 2007. Introduction: Radical conservatism – The Protean Politics of Adat. In *The Revival of Tradition in Indonesian Politics. The Deployment of Adat from Colonialism to Indigenism* (pp. 1–49). London: Routledge.

Hoogerwerf A. 1970. *Ujung Kulon, the Land of the Last Javan Rhinoceros*. Leiden: EJ Brill.

Ibrahim YK, Tshen LT, Westaway KE, Cranbrook E, Humphrey L, Muhammad RF, … & Peng LC. 2013. First discovery of Pleistocene orangutan (*Pongo* sp.) fossils in Peninsular Malaysia: Biogeographic and paleoenvironmental implications. *Journal of Human Evolution* 65(6): 770–797.

Ingold T. 1988. *What Is an Animal?* London: Routledge.

Iskandar E. 2001. Population survey of the Javan gibbon (*Hylobates moloch*) at the Ujung Kulon National Park, West Java, Indonesia. *American Society of Primatologists Bulletin* 25: 9.

Jablonski N. 1998. The response of catarrhine primates to Pleistocene environmental fluctuations in East Asia. *Primates* 39(1): 29–37.

Janson HW. 1952. *Apes and Ape Lore in the Middle Ages and the Renaissance* (Vol. 20). Studies of the Warburg Institute. London: University of London.

Johansson Å. 2015. *Natureculture Origined: An Intersectional Feminist Study of Notions of the Natural, the Healthy and the Palaeolithic Past in the Popular Science Imaginary of Biomechanics*. MA Thesis, Linköping University, Sweden.

Koentjaraningrat. 1985. *Javanese Culture*. Singapore: Oxford University Press.

Laland KN, Sterelny K, Odling-Smee J, Hoppitt W, Uller T. 2011. Cause and effect in biology revisited: Is Mayr's proximate-ultimate dichotomy still useful? *Science* 334(6062): 1512–1516.

Lévi-Strauss C. 1963. *Totemism*, translated by R Needham. Boston, MA: Beacon.

Louys J, Curnoe D, Tong H. 2007. Characteristics of Pleistocene megafauna extinctions in Southeast Asia. *Palaeogeography, Palaeoclimatology, Palaeoecology* 243(1–2): 152–173.

Lowe C. 2004. Making the monkey: How the Togean macaque went from "new form" to "endemic species" in Indonesians' conservation biology. *Cultural Anthropology* 19(4): 491–516.

MacArthur RH, Wilson EO. 1967. *The Theory of Island Biogeography*. Princeton, NJ: Princeton University Press.

Maffi L. 2004. Maintaining and restoring biocultural diversity: The evolution of a role for ethnobiology. *Advances in Economic Botany* 15: 9–35.

Malone NM. 2007. *The Socioecology of the Critically Endangered Javan Gibbon (Hylobates moloch): Assessing the Impact of Anthropogenic Disturbance on Primate Social Systems*. PhD Thesis, University of Oregon, Eugene.

Malone N, Fuentes A. 2009. The ecology and evolution of hylobatid communities: Proximate and ultimate considerations of inter- and intraspecific variation. In Whittaker D, Lappan S. (Eds.), *The Gibbons: New Perspectives on Small Ape Socioecology and Population Biology* (Chp. 12, pp. 241–264). Series Title: Developments in Primatology: Progress and Prospects. New York, NY: Springer Academic Press.

Malone N, Iskandar J, Partasasmita R, Rohmatullayaly EN, Permana S, Iskandar B. 2020. The persistence of silvery gibbons (*Hylobates moloch*) within Ujung Kulon National Park, Banten Province, Java, Indonesia. *American Journal of Physical Anthropology* 171(S69): 172.

Malone N, Ovenden K. 2017. Natureculture. *The International Encyclopedia of Primatology*. Wiley Online Library.

Malone N, Selby M, Longo SB. 2014a. Political and ecological dimensions of silvery gibbon conservation efforts: An endangered ape in (and on) the verge. *International Journal of Sociology* 44(1): 34–53.

Malone N, Wade AH, Fuentes A, Riley EP, Remis MJ, Jost Robinson C. 2014b. Ethnoprimatology: Critical interdisciplinarity and multispecies approaches in anthropology. *Critique of Anthropology* 38: 8–29.

Malone N, Wedana M. 2019. Struggling for socio-ecological resilience: A long-term study of silvery gibbons (*Hylobates moloch*) in the fragmented Sancang Forest Nature Reserve, West Java, Indonesia. In Behie A, Teichroeb J, Malone N. (Eds.), *Primate Research and Conservation in the Anthropocene* (pp. 17–32). Cambridge: Cambridge University Press.

McCarthy JF. 2000. The changing regime: Forest property and Reformasi in Indonesia. *Development and Change* 31(1): 91–129.

McConkey K, Caldecott J, McManus E. 2005. Asia: Indonesia. In Caldecott J, Miles L. (Eds.), *World Atlas of Great Apes and their Conservation* (pp. 417–424). Berkeley: University of California Press.

Megantara EN. 1995. Distribusi, habitat dan populasi Owa (*Hylobates moloch*: Cabrera 1930) di Cagar Alam Leuweung Sancang, Jawa Barat. Bandung: Universitas Padjadjaran.

Meijerink AMJ. 1977. A hydrological reconnaissance survey of the Serayu River basin, Central Java. *ITC Journal* 4: 646–673.

Morphy H, Morphy F. 2006. Tasting the waters: Discriminating identities in the waters of Blue Mud Bay. *Journal of Material Culture* 11: 67–85.

Myers N, Mittermeier RA, Mittermeier CG, da Fonseca GAB, Kent J. 2000. Biodiversity hotspots for conservation priorities. *Nature* 403: 853–858.

Nekaris KAI, Shekelle M, Wirdateti Rode EJ, Nijman V. 2013. *Nycticebus javanicus. The IUCN Red List of Threatened Species* 2013: e.T39761A17971158.

Nijman V. 2001. *Forests (and) Primates: Conservation and Ecology of the Endemic Primates of Java and Borneo.* Wageningen: Tropenbos International.

Nijman V. 2005. *In Full Swing. An Assessment of the Trade in Gibbons and Orang-utans on Java and Bali, Indonesia.* Kuala Lumpur: TRAFFIC South-east Asia.

Nijman V, Nekaris K. 2015. Traditions and trade in slow lorises in Sundanese communities in Southern Java, Indonesia. *Endangered Species Research* 25: 79–88.

Nijman V, Richardson M. 2008. *Presbytis comata. The IUCN Red List of Threatened Species* 2008: e.T18125A7664645.

Nijman V, Supriatna J. 2008. *Trachypithecus auratus. The IUCN Red List of Threatened Species* 2008: e.T22034A9348260.

Oates JF. 1999. *Myth and Reality in the Rain Forest: How Conservation Strategies are Failing in West Africa.* Berkeley: University of California Press.

O'Flaherty WD. 1988. *Other Peoples' Myths.* New York, NY: Macmillan.

Ong P, Richardson M. 2008. *Macaca fascicularis* ssp. *fascicularis. The IUCN Red List of Threatened Species* 2008: e.T39768A10255883.

Permana S, Iskandar J, Parikesit, Husodo T, Megantara EN, Partasasmita R. 2019. Changes of ecological wisdom of Sundanese People on conservation of wild animals: A case study in Upper Cisokan Watershed, West Java, Indonesia. *Biodiversitas* 20: 1284–1293.

Permana S, Partasasmita R, Iskandar J, Rohmatullayaly EN, Iskandar BS, Malone N. 2020. Traditional conservation and human-primate conflict in Ujungjaya Village Community, Ujung Kulon, Banten, Indonesia. *Biodiversitas* 21(2): 521–529.

Pimm SL, Brooks TM. 2000. The sixth extinction: How large, how soon, and where? In Raven PH. (Ed.), *Nature and Human Society: The Quest for a Sustainable World* (pp. 46–62). Washington, DC: National Academy Press.

Ramos AR. 1990. Ethnology Brazilian style. *Cultural Anthropology* 5(4): 452–472.

Reisland MA. 2013. *Conservation in a Sacred Forest: An Integrated Approach to Assessing the Management of a Community-Based Conservation Site.* PhD Thesis, University of Wisconsin, Madison.

Reisland MA, Lambert JE. 2016. Sympatric apes in sacred forests: Shared space and habitat use by humans and endangered Javan gibbons (*Hylobates moloch*). *PLoS ONE* 11(1): e0146891.

Riley EP. 2005. *Ethnoprimatology of Macaca tonkeana: The Interface of Primate Ecology, Human Ecology and Conservation in Lore Lindu National Park, Sulawesi, Indonesia.* PhD Thesis, University of Georgia.

Riley EP. 2007. The human–macaque interface: Conservation implications of current and future overlap and conflict in Lore Lindu National Park, Sulawesi, Indonesia. *American Anthropologist* 109: 473–484.

Riley EP. 2020. *The Promise of Contemporary Primatology.* New York, NY: Routledge.

Seidensticker J. 1987. Bearing witness: Observations on the extinction of *Panthera tigris balica* and *Panthera tigris sondaica.* In Tilson RL, Seal US (Eds.), *Tigers of the World: The Biology, Biopolitics, Management, and Conservation of an Endangered Species* (pp. 1–8). Park Ridge, NJ: Noyes.

Shanklin E. 1985. Sustenance and symbol: Anthropological studies of domesticated animals. *Annual Review of Anthropology* 14: 375–403.

Smith JH, King T, Campbell C, Cheyne SM, Nijman V. 2017. Modelling population viability of three independent Javan gibbon (*Hylobates moloch*) populations on Java, Indonesia. *Folia Primatologica* 88(6): 507–522.

Sponsel LE. 1997. The human niche in Amazonia: Explorations in ethnoprimatology. In Kinzey WG. (Ed.), *New World Primates* (pp. 143–165). New York, NY: Aldine De Gruyter.

Stark M. 1989. Yogyakarta's bird market: Is there an alternative? *Voice of Nature* 76: 42–45.

Stein AB, Athreya V, Gerngross P, Balme G, Henschel P, Karanth U, Miquelle D, Rostro-Garcia S, Kamler JF, Laguardia A, Khorozyan I, Ghoddousi A. 2020. *Panthera pardus* (amended version of 2019 assessment). *The IUCN Red List of Threatened Species* 2020: e.T15954A163991139.

Storm P, Aziz F, de Vos J, Kosasih D, Baskoro S, van den Hoek Ostende LW. 2005. Late Pleistocene *Homo sapiens* in a tropical rainforest fauna in East Java. *Journal of Human Evolution* 49(4): 536–545.

Strang V. 2010. Mapping histories: Cultural landscapes and walkabout methods. In Vaccaro I, Smith EA, Aswani S. (Eds), *Environmental Social Sciences: Methods and Research Design* (pp. 132–156). Cambridge: Cambridge University Press.

Supriatna J, Mootnick A, Andayani N. 2010. Javan gibbon (*Hylobates moloch*): Population and conservation. In Gursky-Doyen S, Supriatna J. (Eds.), *Indonesian Primates* (pp. 57–72). New York, NY: Springer.

Syamsidik B, Luthfi M, Suppasri A, Comfort LK. 2020. The 22 December 2018 Mount Anak Krakatau volcanogenic tsunami on Sunda Strait coasts, Indonesia: Tsunami and damage characteristics. *Natural Hazards and Earth Systems Sciences* 20: 549–565.

Taylor JG. 2003. *Indonesia: Peoples and Histories.* New Haven, CT: Yale University Press.

USAID. 2004. *USAID Strategic Plan for Indonesia 2004–2008: Strengthening a Moderate, Stable, and Productive Indonesia.* Unrestricted Version, published July 28, 2004.

van Balen B. 1994. Insular ecology of forest bird communities of Java and Bali. *Journal of Ornithology* (formerly *Journal für Ornithologie*) 135: 210–211.

van Steenis CGGJ, Schippers-Lammerste AF. 1965. Concise plant-geography of Java. In Backer CA, Bakhuizen van den Brink RC. (Eds.), *Flora of Java* (Vol. 1). Groningen: Noordhoff.

Wallace AR. 1869. *The Malay Archipelago*. Singapore: Periplus Editions.

Wedana APIM, Jeffery S. 2016. Reinforcing the Javan silvery gibbon population in the Mount Tilu Nature Reserve, West Java, Indonesia. Programme of the XXVIth Congress of the International Primatological Society, Chicago, IL, USA, August 21–27, 2016.

Wessing R. 1986. *The Soul of Ambiguity: The Tiger in Southeast Asia*. Monograph Series on South-East Asia, Special Report no. 24. DeKalb: Center for Southeast Asian Studies, Northern Illinois University.

Wessing R. 1993. A change in the forest: Myth and history in West Java. *Journal of Southeast Asian Studies* 24(1): 1–17.

Wessing R. 1994. Which forest? Perceptions of the environment and conservation on Java. *Masyarakat Indonesia* 20(4): 51–68.

Wessing R. 1995. The last tiger in East Java: Symbolic continuity in ecological change. *Asian Folklore Studies* 54: 191–218.

Wessing R. 2006. Symbolic animals in the land between the waters: Markers of place and transition. *Asian Folklore Studies* 65(2): 205–239.

Wetterberg A, Sumataor S, Pritchett L. 1999. A national snapshot of the social impact of Indonesia's crisis. *Bulletin of Indonesian Economic Studies* 35(3): 145–152.

Wheatley B. 1999. *The Sacred Monkeys of Bali*. Prospect Heights, IL: Waveland Press, Inc.

Whitten T, Soeriaatmadja RE, Afiff SA. 1996. *The Ecology of Java and Bali*. Singapore: Periplus Editions, Dalhousie University.

Winchester S. 2004. *Krakatoa: The Day the World Exploded*. London: Penguin UK.

5   Betwixt and between

Apes in (and on) the verge

In 2016, a 17-year-old male, western lowland gorilla named Harambe was shot and killed in Cincinnati, Ohio.[1] The action was taken to protect a four-year-old human child that had fallen into the gorilla enclosure. The news travelled around the world and sparked debates among parents and primatologists alike. Smartphone camera technology allowed for the dramatic moments to be captured, shared and painstakingly analysed. This incident re-kindled long-standing debates about animal rights and the role of zoos in society. Even parenting practices were scrutinised. Half a planet away in Auckland, I found myself discussing the tragic event with my university students in a class on primate behaviour, ecology and conservation. We discussed the behaviour and demeanour of the silverback male, the options available to zoo staff, and the various scenarios by which the child could have been either safely retrieved, further injured or perhaps even killed – with or without harm falling upon Harambe. I presented the following hypothetical situation to my students: imagine it was my child that fell into the gorilla enclosure that day. I, a primatologist with extensive experience studying ape behaviour, approach the zoo staff and somehow calmly attempt to produce an objective, informed read of the situation. Assessing the silverback's behaviour, I convey my wishes on how to proceed to best protect *both* my child and Harambe. I express my desire to not see Harambe harmed, and lobby for a more cautious wait-and-see approach. In spite of all these hypothetical variables, and regardless of my input, I inform my students that the actual outcome of that fateful day is almost inevitable: Harambe, as it so happened, will die by a fatal gunshot wound. Why? Experts have debated the likelihood of alternative outcomes based on the behaviour being exhibited by Harambe in the moments preceding his death. However, with near certainty the lethal gunshot is delivered because zoos in the 21st century are primed to follow their internal, risk-averse protocols. The protocols are in place to maintain the integrity of the zoo's contract with the zoo-going public. The zoo's product, if you will, is the safe display of exotic beasts for recreational and educational purposes. The lives of individual apes (Harambe, in this instance) will almost always be de-prioritised in service to this larger project.[2]

DOI: 10.4324/9780367211349-5

Discussions about displaced apes, and the structures that produce them, are rarely easy. Our hominoid kin are routinely displaced from their habitats, hunted for meat, captured for trade, housed in zoos and enlisted as unwitting participants in our conservation interventions. They are also the subjects of research enquiries by field primatologists. In some settings, such as in biomedical research laboratories, the ethical landscape is relatively clear – a logic of utilitarianism justifies the invasive use of one species for the accrual of health benefits to another species (i.e., humans).[3] However, in other settings, such as in zoological gardens, rehabilitation centres and our "wild" field sites, a reduced clarity surrounding the costs and benefits to individuals and/or species demands a more "sophisticated ethical calculus" (Malone et al. 2010: 782). Such analyses tend to justify risks to individual animals as the "means to an end" – whereby the benefits accrue at an organisational level above that of the individual (e.g., conservation of the species; emphasising the importance of ecological roles) (Malone et al. 2010; Malone & Palmer 2014). The resulting justifications are highly subjective with respect to the values, motivations and positionality of the researcher or practitioner.

In the previous chapter, I began to pivot away from more traditional primatological approaches towards an embrace of ethnoprimatology. For me, an ethnoprimatological approach not only adds anthropological value to primatology by explicitly addressing the human cultural, economic and political aspects of primate ecologies (e.g., Sponsel 1997; Fuentes & Wolf 2002; Fuentes 2012; Waller 2016; Dore et al. 2017; Riley 2020), but also comprises core principles of the dialectical framework. Traditionally, primate studies were carried out under the assumption that primates reside in "pristine" settings where "natural" behavioural phenomena are observed and interpreted through evolutionary lenses. It is increasingly recognised that primate social systems are subject to the influences of past and present anthropogenic alterations, and, accordingly, the methods and frameworks of primate researchers have followed suit (Behie et al. 2019). Theoretical and methodological developments now arm ethnoprimatologists in more holistic investigations of the human–nonhuman primate interface, and facilitate (or indeed oblige) the application of these insights into conservation practice. In this chapter, I'll synthesise an array of research findings, as produced by myself, my students and colleagues, across a variety of research settings in order to further explore humankind's complicated relationships with the apes. Some of the settings elude tidy categorisation, such as "in the wild" or "in captivity". From unhabituated wild populations to highly managed captive groups, and the "gray sites"[4] in between, I'll bring to light conflicts of ethics and efficacy in contemporary primate research and conservation. Here again, the dialectical approach will be deployed in search of a synthesis on topics for which polarising and polemical positions tend to dominate the debates. I'll argue that the amplification of dialectical principles within an ethnoprimatological approach, which is itself firmly grounded in an integrated, critical anthropology, will result in a more effective approach to the multispecies complexes and dynamic interfaces that characterise the anthropocene.

Extending ethnoprimatology

In 2018, a special edition of the *International Journal of Primatology* (guest edited by Professors Erin Riley and Sindhu Radhakrishna) brought together a collection of scholarly research that reaffirmed ethnoprimatology's position within the mainstream of contemporary primatological practice.[5] In her opening essay, Erin Riley documents the "maturation of ethnoprimatology" – from the initial work of Leslie Sponsel and Bruce Wheately at the end of the 20th century, to the sophisticated studies that feature in the edition – and champions its cause.[6] Riley zeroes-in on the foundational core of ethnoprimatology which is: "to recognise humans as part of nature (rather than separate from it), and to interpret primate responses to anthropogenic influences not as aberrant behaviour, but as integral and interesting parts of their ecological strategies" (2018: 708). As such, ethnoprimatology is perfectly situated to elucidate human–alloprimate interfaces, identify factors essential for the mitigation of interspecies conflicts and facilitate the coexistence of people and primates. In approximately 20 years, we've seen a dramatic expansion in the diversity of research approaches, as well as the sites and relational spaces where ethnoprimatology is now practised. The short but productive history of ethnoprimatology demonstrates the richness of the discipline's theoretical and methodological toolkit, and the potential for ethnoprimatology to become a nexus for diverse epistemologies within anthropology (and beyond) to interact in an intellectually robust and engaged manner.[7] Amidst this paradigmatic shift, I see the two critical contributions from ethnoprimatology becoming increasingly relevant to primatological practice: (1) a reconsideration of the *sites of primatological enquiries*; and (2) the incorporation of *ethnographic insight*. Let's delve into these two facets of ethnoprimatology which have the potential, both singularly and in combination, to transform the wider discipline.

Sites of primatological enquiries

Typically, ethnoprimatological studies focus on contact points between people and primates within sympatric distributions. Most commonly, this occurs within forests, or within mosaic habitats (e.g., forest/farm; forest/temple) where the respective activities of human and nonhuman primates produce frequent, interspecific interactions. Within these settings, there are varying degrees of anthropogenic influence on primate lives, including: landscape or dietary modification, hunting and/or sustained attention from local people, tourists or researchers (McKinney 2015; Hanson & Riley 2018). In more anthropogenic environments such as urban areas and agricultural landscapes, humans are likely to modify (albeit unintentionally in some cases) primate behaviour – for example, altering primate diets and foraging patterns through the conversion of forests to farms. Alternatively, humans may explicitly attempt to manage primate behaviour by, for example, driving away crop-feeding primates (Lee 2010) or attracting primates with provisions, such as at Hindu temples (Fuentes

2010; Solomon 2016). Further along the management continuum, tourism sites where primates are provisioned, such as in Japanese monkey parks, reflect an even greater degree of management over primates' lives where nutrition, and consequently ranging patterns, are deliberately modified (Knight 2005). At the extreme end of managed settings are zoos and laboratories, where essentially all aspects of primates' lives are managed by humans (Hosey 2005; Palmer & Malone 2018).

For my own contribution to the aforementioned special edition, I teamed up with my former student (now colleague) Dr Alexandra (Ally) Palmer from Auckland, New Zealand. Back in 2011–2012, I, along with Professor Julie Park (a social anthropologist), supervised Ally's masters-level research at the Auckland Zoo. Ally combined quantitative, ethological observations of orang-utan dyads with ethnographic insight of husbandry practices in an attempt to probe the underlying perspectives of the human caregivers and their orangutan charges. Through a mixed-methods, ethnoprimatological approach, this setting revealed itself to be a mutually shaped, behavioural interface where complex relationships played-out both within and between species (Palmer 2012; Palmer et al. 2015). Fast forward to 2018, and Ally was in the midst of finishing her doctoral research based out of University College London. For her PhD, Ally explored ethical debates in orangutan rehabilitation efforts. Together, for the aforementioned special edition, we wrote about the potential to extend ethnoprimatology's reach by attending to settings where humans manage most aspects of primates' lives, such as zoos, laboratories, sanctuaries and rehabilitation centres – locales that we glossed as "managed settings" (Palmer & Malone 2018). Extending the ethnoprimatological perspective to include managed settings enhances the ability to untangle the complex webs of ideas about primates, and how those culturally mediated ideas shape the lives of managed primates. Moreover, lessons learned in managed settings could be applied to questions of ethics and conservation practice, as increasingly, primate habitats the world over are edging into the "gray" zone between wild and captive contexts. Artificially formed dichotomies were collapsing before our eyes.

Exemplifying the liminal nature of "gray zones" are those spaces occupied by displaced animals undertaking the process of rehabilitation and reintroduction. With the ongoing destruction and fragmentation of habitats, as well as captured animals to support the illegal wildlife trade, the number of displaced apes continues to increase. Correspondingly, there has been a recent prioritisation by conservation managers of integrated rehabilitation and reintroduction programmes. Although significant variation exists in the structure and philosophies of these programmes and their associated sites, one commonality is an intentional (and sometimes unintentional) lack of clarity regarding the distinction between "captivity" and "wild" (Palmer & Malone 2018; Palmer 2020). For example, centres, where primates are housed while they undertake the process of rehabilitation, typically resemble zoo settings insofar as humans, by-and-large, control primate diets, social interactions, housing and breeding. Here, the exposure to human contact and management practices are highly variable, but a common goal is for some animals to regain self-sufficiency (as much as possible within the confines of the centre) and break any bonds that may have formed with humans. For some animals, the process of rehabilitation may be deemed "successful", thus qualifying them as candidates for reintroduction. However, even the act of release is by no means the end of human management, as there are varying degrees of intervention at release sites. For example, at many ape reintroduction sites, supplemental food is provisioned to both aid in the transition to a wild diet, assist in the post-release monitoring of released apes and facilitate any required medical interventions (Palmer & Malone 2018).

Let me provide an example here to demonstrate the point. Between 2011 and 2016 more than 30 silvery gibbons have been confiscated from private ownership by the Javan Primates Rehabilitation Centre (JPRC), acting with the support of an overseas funder (Aspinall Foundation) and government agencies.[8] After a period of quarantine, the gibbons, depending upon their age and health status, are cared-for by the veterinarian and husbandry staff at JPRC in Ciwidey, West Java. An underlying motivation for this project is to re-establish a viable, self-sustaining population of silvery gibbons in the protected Mount Tilu Nature Reserve. To this end, gibbons are socialised in a variety of group compositions and enclosure arrangements, but an emphasis is placed on the formation of bonded, male-female pairs. In 2014–2015, three groups of rehabilitated gibbons were reintroduced to the forest at a suitable release site within the Nature Reserve (Wedana Adi Putra & Jeffery 2016). The post-release story of one pair is of particular interest. Adult male *Cheri* and adult female *Ukong* were paired together at JPRC, and over the course of two years had formed a stable bond. In fact, in July 2013 *Ukong* gave birth an infant (*Uchi*) at the centre. In May 2014, the three gibbons were released into the reserve, and a period of post-release monitoring commenced. Within weeks of release, 10-month-old *Uchi* had disappeared.[9] Shortly thereafter, a behaviourally and nutritionally distressed *Ukong* required evacuation, and was transported back to the rehabilitation centre. Meanwhile, a wild female gibbon repeatedly approached the male *Cheri*, and within a period of four months they were exhibiting behaviour associated with the formation of a new pair bond. The ability to observe these processes were enabled through the intermittent use of an arboreal feeding station (Figure 5.1). While the continued "management" of the once-wild/once-captive *Cheri* was planned, the use of the feeding station by the wild female (later named *Dewi*) was an unanticipated extension of the "managed" site. At last report, the newly formed pair remained stable, and are confirmed to have reproduced (M. Wedana Adi Putra, personal communication).

In my experience with various projects in Indonesia, an increasing number of apes find themselves in settings that are "managed", in some way, by humans. From the blurring of boundaries between "captive" rehabilitation centres and "wild" reintroduction sites, to the gibbons "passively habituated" by spiritual pilgrims at the Leuweung Sancang nature reserve (see Chapter 4), in-situ and ex-situ primate populations can "affect each other in reciprocal ways"

Figure 5.1 Left: a rehabilitated and reintroduced male silvery gibbon (*Cheri*) accessing a temporary feeding station at a release site. Right: the wild female (*Dewi*) also learned to access the feeding station. After his release partner (*Ukong*) required evacuation and a return to the rehabilitation centre, *Cheri* and *Dewi* became a stable pair.

Photo credits: Nicholas Malone.

(Braverman 2014: 53). As in the above example, primates at rehabilitation centres are typically displaced from free-ranging populations, but can potentially be returned to the wild to either interbreed with existing populations or form new populations, though interbreeding with existing populations remains controversial (Smits et al. 1995; Rijksen & Meijaard 1999; Rosen & Buyers 2002; Beck et al. 2007; Russon 2009; FORINA 2013). In a case that exemplifies the fluid movement of individuals, a female silvery gibbon named *Regina* – originally born in, and transferred between, Australian zoos before being sent to the United Kingdom – was "repatriated" to the JPRC in West Java. There, she was paired with a displaced wild gibbon (*Aom*), and produced an offspring in the centre (*Rega*). Subsequently, the three gibbons were released into the Mount Tilu release site (Wedana & Jeffery 2016). This fluid movement of individuals is underpinned by practices and philosophies of management – namely captive breeding in the name of conservation. While, historically, primates in zoos were removed from their original habitats, today's progressive zoos refrain from this practice and instead sustain captive populations through managed breeding programmes. Progressive zoos emphasise not only their role as conservation educators but also as "arks" harbouring threatened animals, some of whom (ostensibly) may be required to repopulate the wild (Oates 1999; Palmer & Malone 2018). Captive breeding in zoos – followed by release – has proved moderately successful for some primates, most famously golden lion tamarins (*Leontopithecus rosalia*: Kleiman et al. 1991). Similar interventions have recently been conducted with western gorillas (*Gorilla gorilla gorilla*: King & Courage 2008) and Sumatran orangutans (*P. abelii*: Cocks & Bullo 2008; Bullo 2015) producing mixed results. These conservation tactics represent an expensive gamble. Striking a note of caution, Oates (1999) warns against the potential of captive breeding programmes to divert resources from protectionist interventions. In essence, Oates' position represents a more traditional delineation between "in-situ" and "ex-situ" conservation:

> it is probably wise to ensure that small breeding colonies of at least the most threatened animal species are held in captivity as a safeguard against the failure of conservation in the natural habitat. But captive breeding in zoos should not be regarded as the primary means by which endangered animals may be conserved, and the maintenance of captive-breeding colonies must not be allowed to distract attention from the highest conservation priority: the protection of wild animals in their original habitats.
>
> (Oates 1999: 228)

However, I (and others) argue that such unambiguous distinctions between wild and captive populations are becoming increasingly tenuous. For better or for worse, the liminal "gray sites", where reciprocal influences both within and between species are managed, represent new sites of primatological relevance.

Ethnographic insight

As discussed throughout the book, conceptualising the entanglement of human and alloprimate lives is becoming common practice within primate research and conservation. As such, ethnoprimatology is a central locale for the growing consensus in primate studies that the interface between humans and other primates is more than one of person/subject, and that understanding relationships between researchers and their nonhuman study subjects can be an essential element in primatological practice (Asquith 2011; Malone et al. 2014). While all ethnoprimatological studies include a consideration of the human dimensions within a human–alloprimate interface, not all are designed to capture the inner realms of meaning and intentionality behind human behaviour. To obtain the latter, or true ethnographic insight, a/an primatologist/anthropologist must commit considerable time to the endeavour throughout all phases of the research. In terms of preparation, one should seek the relevant training in sociocultural theory, methods and ethics, and become at least moderately proficient in the language spoken by the people who are expected to become research participants, interlocutors and collaborators. In the field, time must be budgeted for the "mutual familiarisation between researcher and informants to take place such that the latter are willing to share their perspectives, experiences and knowledge with the former" (Knight 2017: 175). All of this, of course, must be balanced with the time required to study the salient behavioural and/ or ecological aspects of the nonhuman primate(s) of interest. Daunting as it sounds, in many ways it is part and parcel of conducting anthropological fieldwork, generally speaking. Many of us (primatologists) may already be primed to adopt ethnographic principles into our research (i.e., by routinely seeing our study animals as individuals and recognising their subjectivities) (Riley 2020).

At a time when some primatologists are embracing approaches that merge ethnographic engagement with primate studies, multispecies ethnographies are becoming more common in social anthropology (e.g., Haraway 2008; Kirksey & Helmreich 2010; Ingold 2012; Ogden et al. 2013). An influential contribution comes from Kirksey and Helmreich (2010) whereby animals "previously appearing on the margins of anthropology – as part of the landscape, as food for humans, as symbols – have been pressed into the foreground" (Kopnina 2017: 338). Notions of "interdependencies" and "webs of life" are prominent in these conceptualisations. Bringing the focus back around to fieldwork, a hallmark of anthropological practice, Alan Smart (2014: 4) succinctly connects the historical and contemporary leanings of the discipline, stating:

> An insistence on the utility of fieldwork in illuminating the interaction between different species links us back to the pre-postmodern moment in anthropology when animals and plants were central elements of ethnographic research, when cattle, coral gardens, rice cultivation and so on were thoroughly embedded and attended to in almost every major anthropological monograph.

In these endeavours the distinction between "human worlds" and "nature" is discarded and multispecies entanglements become central aspects of anthropogenic ecologies (Malone et al. 2014: 8). Anthropologist Alex Nading (2014: 11) defines "entanglement" as: "the unfolding, often incidental attachments and affinities, antagonisms and animosities that bring people, non-human animals and things into each other's worlds". Here again, ethnoprimatology provides not only a suitable vehicle to operationalise the concept of entanglements, but also to move past historical rifts in anthropological practice (Riley 2006, 2013). Two critical elements of this perspective that resonate with dialectical principles are: 1) a commitment to materialist understandings of the nonhuman contexts salient to human existence; and 2) ethnographic fieldwork's inherent ability to temper the tendency for theorising to outpace the accumulation of data (Smart 2014). For primate research and conservation in the anthropocene, adopting ethnographic methods has been an important development in the move towards attributing symbolic, social and ecological agency to the actors within a given human–alloprimate interface.

In 2014, I was fortunate to take the lead on a multi-authored publication in the journal *Critique of Anthropology*. As a collective of both junior and senior scholars, we sought to demonstrate the ways in which ethnoprimatological approaches are firmly situated within an integrated, critical anthropology (Malone et al. 2014).[10] We put forward three case studies in an effort to identify unifying themes and effective anthropological approaches to the dynamic, multispecies interfaces that are characteristic of the anthropocene. The first case study featured the work of Alison Wade, then a masters-level research student from the University of Auckland. As previously mentioned, given the large numbers of displaced apes, there has been a recent prioritisation of integrated rehabilitation and reintroduction programmes. In addition to determining the suitability of the habitat for the release of rehabilitated primates, IUCN guidelines also require an assessment of the local residents' attitudes towards the conservation initiative (Baker 2002). At the time, the JPRC was interested in the feasibility of the adjacent Mount Tikukur as a potential release site. Alison's research began an effort to understand why activities such as forest clearing and encroachment into protected areas occur, as well as factors that might lead to the hunting and/or capture of released primates. Alison, with the assistance of myself and local collaborators, conducted cultural mapping exercises and semi-structured interviews. Participants were recruited into the study using a purposive sampling method to identify individuals whose livelihoods draw upon resources in the boundaries between forest and farmland (Tongco 2007; Bernard 2011) (Figure 5.2).

Participants of Sundanese ethnicity came from several villages within the district of Patuhua, Ciwidey. Participants were accompanied on a typical trip to the forest, taking note of areas they identified as important. Semi-structured interviews were used to build upon the information gained from the forest mapping exercise. Land ownership/management in the area consists of land that is either privately owned or leased from the government, as well as demarcated blocks of both protection and production forest. Directly adjacent to protection and production forests is a network of hillside terraces and channels that ensure each farm has an adequate supply and distribution of water to irrigate their crops. The main crop grown in the area is strawberry, which is actively marketed to local tourists. In addition to strawberries, other crops include rice, potatoes, peppers, tomatoes, tea, spring onion and cabbage. The majority of the water in this area is sourced from a mountain spring that is collectively managed by the people. The cultural mapping and interview

Figure 5.2 Then graduate student (now PhD, Auckland 2020), Alison Wade during her research to assess local attitudes towards conservation initiatives in West Java. Alison is accompanied by long-term field assistant Ajat Surtaja (right).
Photo credit: Nicholas Malone.

process revealed that consistent access to water, as well as its fair distribution, was the primary concern of the local residents. Its importance was voiced by every participant, usually in relation to questions about conservation or their perceptions of the forest. One participant clearly illustrated this viewpoint in articulating the importance of "... *penghijauan* [reforestation] to conserve water resources. No forest means no water because the roots of the trees preserve water". Additionally, the quality of the water is particularly important, and this was of concern to local residents in regards to the establishment of the JPRC since one of the main water channels runs through land leased by the primate rehabilitation project. In addition to the necessity of water for farming, one participant stated: "to conserve our nature is everybody's obligation for everybody's benefit. If there is no animal we feel inappropriate because we live together side-by-side with animals". A similar response was given by another participant, who said animals should not be hunted on account of the aesthetic value they provide "as ornaments or decorations of the forest". Two participants shared the notion that the animals protect the forest and this is why they should be conserved. Of the participants interviewed in this research the main reason given as to why the forest is important was to conserve their water resources, followed by the protection it offers them from landslides (*tanah longsor*), and finally the daily needs it provides in terms of food and wood. The participants had an understanding of the term "conservation", although their main focus was to conserve the forest to protect water resources. Only two participants extended the concept of conservation to the animals that reside in the forest (Wade 2012; Malone et al. 2014: 12–13).

In the second case study of the 2014 *Critique of Anthropology* article, Erin Riley shared her insights from years of ethnoprimatological research in Sulawesi, Indonesia. In this research, a diverse methodological toolkit was deployed to explore the expanded ecological community comprising humans and Tonkean macaques (*Macaca tonkeana*) in Lore Lindu National Park. Through a combination of primatological, ecological and ethnographic research methods, Riley describes a complex pattern of human–macaque relationships. In Lindu, diverse human experiences (e.g., migrant vs. indigenous histories; knowledge of the forest and encounters with its denizens; perceptions and experience of crops loss due to macaque foraging) are related to a range of conservation-related perspectives (e.g., utilitarian vs. intrinsic valuations of the forest). Furthermore, Riley's ethnographic research reveals salient connections between indigenous Lindu and macaques, as manifested in folklore, that were not present among more recent migrants (Riley 2010; Riley & Priston 2010). In the third case study presented in Malone et al. (2014), Melissa Remis and Carolyn Jost Robinson present a summary of a multifaceted research programme exploring a myriad of human–wildlife relationships within the Dzanga Sangha region of the Central African Republic. Remis and Jost Robinson document trends in wildlife abundance and behaviour, while simultaneously engaging with human hunting practices, prey preferences and attitudes towards wildlife. Through a combination of hunter off-take surveys, interviews and more traditional primatological methods, Remis and Jost Robinson elucidate a level of nuance in human–wildlife relationships that would be concealed in more narrowly framed investigations of either people or animals. In this publication and elsewhere (e.g., Jost Robinson 2012; Remis & Jost Robinson 2012), such studies "provide a window into the ways that humans value animals and how these relationships are (re)negotiated in light of wildlife decline" (Malone et al. 2014: 19).

The three distinct, but overlapping, case studies demonstrate the shifting nature of primatological research in the 21st century. Through ethnographic insight, the emergence of several themes is evident, including: multiple meanings and implications for the concept of value; ecological knowledge and the processes that change communities over time (e.g., migration); and the sensitivity of conservation awareness to personal histories and experiences. Here, I will briefly touch-on the concept of value. When primatologists embark upon the study a focal primate species (or a community of species), the sociocultural and political contexts, and implications of these endeavours matter greatly. For Remis and Jost Robinson, capturing the intrinsic and extrinsic valuation of wildlife by various stakeholders is facilitated by the "transvalued species concept" (Remis & Hardin 2009). The transvalued species concept connects the economic, ecological and symbolic valuations of wildlife species across cultural, economic and geographical boundaries. For example, gorillas appear prominently in western as well as in BaAka peoples' (within the Dzanga Sangha region) folk tales and are personified in both cultures. Encounters with gorillas can provide monetary benefits but regardless are significant and often life-altering for tourists, researchers and hunters alike. In other contexts, the granting of scientific and economic significance to research subjects within the forest can result in an enhanced perception of extractable value from the forest (Malone et al. 2010). While often considered an essential component of conservation in the 21st century (i.e., ascribing a higher value to biodiversity protection vs. biodiversity exploitation), I suggest the potential for negative and/or unpredicted repercussions (e.g., increasing the threat of capture for an illegal pet trade that profits from the perception of rarity and endangered species status) (Malone et al. 2004; Nijman 2005; Courchamp et al. 2006; Siriwat et al. 2019). Might, for example, the promotion of scientific/ecological value produce short-term results but risk a wider attraction by those wishing to extract the value that has been added? To this final point, one participant from Java spoke forebodingly about the potential gibbon release site and warned that: "people from outside the area would be (drawn into) hunting the animals in the forest". This specific example demonstrates a broader theme of the book: attention must be paid to the dialectical relationship between agents in a complex system. Emphasising the connections between different ways of valuing wildlife and integrating social as well as ecological research can reinvigorate global and local populations, providing one potential way forward for conservation. Strict protectionism and enforcement of laws protecting endangered wildlife may be another. Either way, generous engagements with anthropological theory and methods, including healthy doses of reflexivity (both personal and epistemo-logical), are becoming essential to the efficacious conduct of primate research and conservation.[11]

Ethical entanglements

If, as according to my argument, the future of primatological practice includes an expanded notion of the *sites of primatological relevance* and an increasing incorp-oration of *ethnographic insight*, then an explicit engagement with the relevant *ethical implications*, for humans and nonhumans alike, takes on heightened sig-nificance. Already, an increase in attention is being paid to the ethical challenges faced by primatologists (Fedigan 2010; MacKinnon & Riley 2010; Malone et al. 2010; MacKinnon & Riley 2013; Riley & Bezanson 2018). These discussions begin with painting a more realistic picture of the multitude of complex, eth-ical issues faced by field primatologists. Even during so-called "noninvasive research", primatologists routinely encounter a whole host of ethical issues that involve not only the alloprimates under study (e.g., risk of disease trans-mission; an increased susceptibility to hunting following a period of habitu-ation), but also the people whose livelihoods are intricately linked to either our study animals or our study sites (McLennan & Hill 2013). In response, a more proactive approach for interacting with animal ethics committees is being promoted (Fedigan 2010; MacKinnon & Riley 2013). Elsewhere, I, along with Ally Palmer, have argued for an extension of this perspective to form an "ethic-ally grounded primatology", whereby researchers are compelled to: (a) consider the impact of their work on individual study animals and broader ecosystems; (b) recognise that personal values may diverge from those of local people; and (c) account for social and political contexts when determining the impacts of their research (Malone & Palmer 2014). To further flesh-out what the con-cept of an ethically grounded primatology means for a dialectical primatologist, I will provide a few specific examples from which more general lessons can be gleaned.

The first example comes via a collaboration with another former stu-dent at the University of Auckland, Dr Courtney Addison (PhD, University of Copenhagen), who is now a lecturer in the Centre for Science in Society at Victoria University of Wellington in New Zealand. In 2018, Courtney invited me to co-author a response to Edwards et al.'s (2018) controversial article published in *The American Journal of Bioethics* which proposed testing Ebola vaccines in wild apes (specifically chimpanzees). Edwards and colleagues suggest that certain nonhuman primates represent an advantageous research population: first, they also suffer from Ebola and therefore stand to benefit from a vaccine; and, second, they are as good a proxy for humans as we are likely to find. The authors argue from a One Health[12] perspective to infectious dis-ease mitigation that: (a) field trials could bolster the evidence base for Ebola vaccines' efficacy in human subjects, and thereby accelerate licensing; (b) min-imise interspecies transmission; and (c) protect nonhuman primate populations.

However, there are both ethical and scientific issues with attempting to leverage data from a field test for human purposes. Scientifically, the transference of insight from animal models (including nonhuman primates) to human biology is unclear. While chimpanzees have historically stood in as models of human biology due to these species' observable similarities, Jones and Greek (2014) convincingly demonstrate that complex, systemic differences between species consistently prove to be more biologically meaningful. Ethically, the rationale for testing an unproven medical product on an endangered species for the benefit of a(n) (overly) populous one is problematic. This is especially so when the latter group (humans) can consent and meaningfully shape research, while the former (a nonhuman species) cannot. In other words, research can be done *with* human subjects, but only *on* nonhuman subjects (Addison & Malone 2018: 53, emphasis original). We also note the "moral inconsistency" of treating chimpanzees as similar enough to humans to serve as useful model organisms, but different enough to subject to invasive testing (Jones & Greek 2014: 494). Finally, testing unproven vaccines on wild populations of chimpanzees (or other apes) exposes them to potential adverse effects that may arise from the medical product or its administration. These populations already face a host of other risks, including hunting, habitat loss and other infectious diseases. The risks of a field trial add to, and potentially compound, pre-existing risks. We must assess how these risk factors interact to impact endangered ape populations, and whether a field trial might produce a combined burden greater than each risk taken independently (Addison & Malone 2018: 54). Thus, I contend that potential human health benefits are not a strong rationale for testing novel vaccines on endangered primates, and such trials should only be justified with respect to the species that bears the cost. Until that time, any progression towards the types of experiments advocated by Edwards and colleagues should be met with swift and strong opposition from the primatological community.[13]

A second example comes from the doctoral research of Alison Wade who conducted a multifaceted, ethnoprimatological study of the human–ape interface in Cameroon (Wade et al. 2019; Wade 2020). In Cameroon, a significant proportion of both the Cross River gorilla (*Gorilla gorilla diehli*) and the Nigeria-Cameroon chimpanzee (*Pan troglodytes ellioti*) population range across landscapes with no formal protective status (Morgan et al. 2011; Bergl et al. 2012). An understanding of these apes' distribution across the landscape is required in order to tailor conservation approaches to areas that have no formal legal protection. To do so, the establishment of strong relationships with the local communities who are highly dependent on forest resources for their livelihoods and subsistence needs (e.g., food, medicine and materials for housing construction) is paramount. Conservation action plans for these apes acknowledge the need to maintain local customs and traditions that facilitate wildlife conservation. They also emphasise the need to improve educational outreach programmes in an effort to foster greater local responsibility for forest and wildlife conservation (Morgan et al. 2011; Dunn et al. 2014). These recommendations are based on the assumption that education will increase the rate of change in societal attitudes towards wildlife, and that positive attitudes will persist in the long term (Jacobson 2010). However, there is limited knowledge surrounding how local communities perceive the apes or the forests within Cameroon. Through the combined use of botanical surveys, analysis of nesting sites, participant observation, semi-structured interviews and archival research, Alison's research obtained nuanced insight into the human–ape interface in an unprotected landscape with entangled social and ecological histories.[14]

Alison found that chimpanzees and gorillas had distinct nesting ranges within the Mone-Oku Forest. While both taxa are threatened by habitat alteration, they are susceptible in different ways. The Mone-Oku chimpanzee nesting range is not directly threatened, but there is the potential for the loss of forest connectivity and isolation of the chimpanzee community. In contrast, parts of the Mone-Oku gorillas nesting range is directly threatened by habitat loss, especially favoured areas of lowland and secondary forest. There is also the possibility of future interspecies interactions between the gorillas and villages as farms encroach into the gorillas' range (Wade et al. 2019). There are seven villages within the area that access the forest and its resources, with Eshobi and Nga claiming traditional ownership over the study area. Alison found that participants from Nga and Eshobi villages held contradictory ideas and context-dependent perceptions of the apes. However, the most important issue discovered was the near-universal perception of the Mone-Oku Forest as an unlimited economic resource. While the Cross River gorillas and Nigeria-Cameroon chimpanzees are primates of special significance to researchers and conservationists, for the villagers of Nga and Eshobi the significance lies in the forest itself rather than particular animals within (Wade et al. 2019; Wade 2020). Interviewees expressed conflicted feelings towards the apes, which varied depending on context. For example, many people feared the apes, but they also recognised the economic benefits the apes brought to their villages through conservation initiatives (e.g., the Gorilla Guardian programme) or the occasional researcher.[15] The level of fear attributed to either the gorillas or chimpanzees differed based on the varied behaviour of the apes in an encounter and people's perceptions of the apes. Wade (2020: 243) summarised the situation succinctly, stating:

> Humans do appear to shape both the chimpanzee and gorilla choice of nesting locations but in multiple ways. The Mone-Oku chimpanzees had a clear preference to nest in mid-elevation closed-canopy forest and on steep slopes that concealed their nests from the main forest roads. The chimpanzees did show preferences towards particular tree species when constructing their night nests, but they also nested in a wide variety of tree species. Overall, the chimpanzees' nesting preferences suggest they are avoiding areas of high anthropogenic activity. Together with the cryptic behaviours we observed indicates they have adapted to a long history of human hunting pressure (Isaac & Cowlishaw 2004) … The Mone-Oku gorillas also nested in areas that were effectively concealed from the main

forest roads and on steep slopes that were difficult to reach. But these nest sites were generally located in areas of high human activity that poses an indirect threat, especially during the rainy season, when traps and snares were placed for other forest animals. While gorillas appear to avoid direct encounters with humans, humans have modified the environment in a way that presently attracts and benefits the gorillas with the secondary forest regrowth occurring in large areas of abandoned farmland.

As clearly evidenced in this example, people are actors within ecosystems and the links between ecology, policy and economy are multidimensional and dialectical (Gezon 2006). The methods and insights gained from an ethnoprimatological approach highlight the complexity and heterogeneity that exists among human communities (Ellwanger et al. 2017). How local communities interact with the apes can change depending on the context at that particular point in time (Wade et al. 2019).

Alison Wade's research elucidates local social structures and traditions, as well as the wider contexts that shape forest activities. To gain these valuable ethnographic perspectives, a researcher must be willing to invest considerable time and energy, and engage in careful and consistent consideration of inter-personal relationships and responsibilities. In short, the ethical ramifications are elevated, and negotiating this terrain may exceed the formal training of some primatologists. In addition to oversight by animal ethics committees, ethnoprimatologists wishing to incorporate an ethnographic perspective must seek approval from Human Participant Ethics Committees or HPECs (as they are in Australia and New Zealand). Social scientists from the United States and the United Kingdom will know them as Institutional Review Boards (IRBs) and University Research Ethics Committees (URECs), respectively. While compliance is typically compulsory, the conventions and constraints of insti-tutional ethics panels present real limitations to their effectiveness (MacClancy 2013). Despite these limitations, a critical engagement with HPEC approval processes can facilitate a more ethical navigation of the human–nonhuman primate interface, especially for new or early-career investigators. While the objectives of Alison's research (or my own, for that matter) have never included the reporting of illegal behaviour, or specifically identifying illegal actors, a common exchange with HPECs involves outlining a response to the witnessing of activities deemed illegal by national laws. For example, the national laws of Cameroon prohibit the killing, capture and/or trade of the focal primate species.[16] However, it is important to note three additional points: (1) com-plex systems of customary rules of behaviour are often given precedence to authoritative structures of the state (especially in rural contexts), thereby com-plicating the researcher's role; (2) a lack of resources at the village level may impact the enforcement of national laws; and (3) previous experiences in the research setting should factor into any decision-making. For example, know-ledge of corruption in authoritative circles often leads to anti-ethical or further illegal handling of initial offences and should be avoided. Mandatory reporting to agencies that have approved the research agenda may present a researcher with an especially challenging set of circumstances. Therefore, an approach that is sensitive to these complexities and takes seriously the guidance of local collaborators is warranted and places a premium on the researcher's informed judgment (Malone et al. 2017). As a matter of practice, any sensitive information shared by participants should remain confidential, and anonymity should be preserved to the best of one's ability throughout the research process. For example, while Alison did not witness any illegal hunting in Cameroon, one participant expressed concern during an interview that chimpanzees had been caught in his traps, but had managed to escape. He did not feel comfortable discussing this with the local wildlife organisation and was concerned about the legal implications. Having a clear understanding of the wildlife laws in Cameroon, Alison was able to advise him of the legal ramifications and suggest he move his traps to areas outside of the chimpanzees' range.

The first two examples in this section on ethical entanglements come from wild settings where various management interventions are either being proposed or enacted. Whether in the form of a proposed experimental trial or a protected habitat corridor, we can see that these specific human–ape relationships are co-produced and influenced by both material and the conceptual factors, placing nonhuman primates in the verges of nature and society. The third and final example comes from an unequivocally "managed setting" – that of the zoo-logical garden. Here, I will be drawing on previous analyses of ethical issues in zoo settings (Malone et al. 2010; Palmer 2012; Malone & Palmer 2014; Palmer et al. 2015, 2016), and conclude by circling back to the discussion of Harambe's death that began this chapter.

Similar to wild or "field" contexts, the zoo poses a number of ethical challenges with respect to engagements with both people and nonhuman primates. Unlike the field primatologist, the researcher in the zoo is but one of many humans observing and interacting with zoo-housed primates. Hosey (2005) suggests that the zoo is distinguished from other nonhuman primate habitats by the regular presence of large numbers of unfamiliar humans, space restrictions and the fact of "being managed". Although other habitats display some of these features (e.g., primate pets face space restrictions, temple-living monkeys are similarly exposed to large numbers of unfamiliar humans), Hosey argues that these three features act in concert to create a distinctive zoological garden environment, which influences primate behavioural repertoires and social interactions in specific ways (Malone & Palmer 2014: 28–29). For these reasons, the ethical issues with the zoo apply less specifically to the researcher, and rather encompass the complexities and problems represented by the zoo institution as a whole.

Recall that in 2012, Ally Palmer studied the relationships between keepers and orangutans at Auckland Zoo using a combination of ethology, discourse analysis of contemporary and historical texts, and ethnography – specifically, participant–observation and interviews. As in our other research settings, the HPEC required substantial documentation outlining how Ally would handle ethical issues relating to informed consent, voluntary participation, confidentiality and anonymity, participants' right of review and the impacts of the research on work positions and practices. In particular, the HPEC asked Ally to ensure that participants would not feel pressured by their colleagues to participate in the research, or that their work performances were being evaluated. In response to these concerns, Ally sought assurances from the Auckland Zoo Director that participants' employment positions would not be affected by either their decisions regarding participation in the research or by any published products of the research. A more complicated ethical concern relating to the observation of work practices was the potential that Ally might observe the mistreatment of orangutans, which could constitute a violation of New Zealand's Animal Welfare Act (Parliamentary Counsel Office 1999).[17] Ally decided that a sound way to balance such concerns would be to report any mistreatment of the orangutans to the Zoo Director, but to first inform any implicated people of her intention to go forward so that they might have a chance to explain themselves. Fortunately for everyone involved, this ethical dilemma remained purely hypothetical, since all Auckland Zoo primate keepers treated the orangutans with great respect and care throughout Ally's fieldwork (Palmer 2012; Malone et al. 2017).

A predominate theme from Ally's research emerged in response to the moral conflict of keeping great apes in captivity. Several of her research participants expressed discomfort with some aspects of zoos (such as cages and the restriction of animals' freedom) but suggested that we need them in order to help protect nonhuman species from imminent extinction (Palmer 2012; Palmer et al. 2016). The educational role of zoos was also sometimes invoked as a reason for keeping great apes in captivity. Husbandry staff employed a somewhat pragmatic approach, noting that "these are the cards they [zoo animals] were dealt", and it makes a large difference morally that they are zoo-bred rather than "ripped out of the wild". Caregivers suggested that zoos might as well make good use of captive great apes as "martyrs" or "ambassadors for their species", since they expressed the view that progressive zoos can do a great deal of good in educating the public about conservation issues. Thus, they argued that zoo animals fulfil important roles of advocacy, which will benefit their entire species even if it means that individually the animals must lead restricted lives (Palmer et al. 2016: 240–241). Bulbeck (2005) suggests that this idea is central to many modern zoos' justifications for their existence. Zoos present their animals "as either sacrificial ambassadors for their kin in the wild or as the last hope for an endangered species" (Bulbeck 2005: 23).

As with research using humans, studies with nonhuman primates (and other animals) also face oversight by institutional ethics committees. Here, I'll refer to these institutional bodies as animal ethics committees (AECs). Ethical bodies that oversee research with human participants tend to consider a comprehensive range of ethical issues, from participant safety to informed consent, anonymity and confidentiality. In contrast, many AECs appeal, somewhat narrowly, to the following principles, commonly known as the "3Rs": replacement (using alternatives to animals whenever possible); reduction (using as few animals as possible); and refinement (minimising distress and suffering) (Russell & Burch 1959). Some AECs go so far as to permit researchers to identify their projects as "exempt" from the entire oversight process, if they consider their research to be "noninvasive" or "purely observational". However, the ethical issues of conducting even noninvasive research on primates in zoos extends beyond the particulars of any given research project. A number of authors have taken issue with the philosophical justification and implications of the act of watching nonhumans in the zoo, and have suggested that the zoo problematically presents nonhumans as subjects of human domination and control. Berger (1980), Malamud (1998) and Acampora (2005) have pointed towards an unequal dynamic between those who look and those who are looked at, which renders animals "absolutely marginal" within the zoo environment (Berger 1980: 24). These authors propose that this "exploitative consumption of an objectified animal, debases the watcher as well as the watched" (Malamud 1998: 5), such that zoos both reinforce and reflect an unhealthy relationship between humans and other animals, which is based on power and exploitation (Malone & Palmer 2014: 29–30).[18]

Finally, returning to Harambe's life (and death) at the Cincinnati Zoo reveals the very crux of the problem: which lives and whose bodies are recognised and valued? In the wake of Harambe's death, Bron Taylor, a professor of environmental studies and ethics at the University of Florida, provocatively wrote:

> Some environmental philosophers and scientists, however, contend that an individual member of an endangered species is more valuable than an individual human being … the value of an individual decreases proportionately with the size of its population. Such arguments are premised upon an understanding that the viability of a species is associated with the variety of genes in its population: With few exceptions, the greater its genetic diversity the greater will be a species' resilience in the face of diseases or environmental threats. But the smaller the population is, the higher is the risk of extinction. Consequently, every individual matters. So, if one starts from an ethical claim that humanity ought not drive other species off the planet, and add scientific understandings about the value of an individual organism to the viability of its species, an endangered animal such as Harambe could be considered more valuable than one that is not valuable in this way.[19]

Or as Helen Kopnina (2017: 341–342) points out:

> If the dual category of human and nonhuman is diffused, shooting a gorilla and saving a toddler becomes problematic. And so does keeping apes in zoos in the first place. Should all animals be treated equally? When the choice between a human and a member of the endangered species is made, what protocol needs to be followed? … Obviously, an answer to this

question is subject to value judgments – of who is accorded rights, and by which human advocates – since nonhumans cannot argue their own case. Unless multispecies ethnography is willing to engage with such questions, it is likely to remain apolitical, without realising the exploitive nature of human–nonhuman relationship[s].

For dialectical primatologists, it is our time to take up Kopnina's challenge, and to question some of the underlying premises of our discipline. From the first scientific description of an anthropoid ape by the Dutch physician Nicolaes Tulp, to Edward Tyson's (1699) examination and dissection of a chimpanzee, the closest relatives to the human species – under the guise of science – have become subjects of knowledge and have been subjected to a separate standard of valuation (Corbey & Theunissen 1995). Such early delineation between human and "other" animal life forms confined the recognition of higher moral status neatly within our own species boundary. Given the etymology of "ethics" [Greek, *ēthos*], I contend that we must begin by questioning "the basic, underlying assumptions, guiding beliefs, and binding customs of our common practices and codes of conduct" in primatological practice (Malone et al. 2010: 780). Although I am unable to offer a definitive resolution to the debate surrounding Harambe's death, I can propose a synthetic position to the conflict at hand (in keeping with the dialectical tradition). That is: although I am sympathetic to the obvious ethical abuses to the rights of individual zoo animals, and give the benefit of the doubt to individual keepers' moral intentions, I do *doubt the benefit* of zoological gardens with respect to the overall representation of an ethical treatment of animals in society at large.

Towards a coexistence with species in (and on) the verge

Be it by theoretical advancements or in response to the pending extinction crisis, primate research and conservation are undergoing a paradigmatic shift. In this chapter, I have addressed some of the core elements of this movement. I have also raised some difficult, ethical questions. I have demonstrated that *the sites of primatological relevance* are increasingly those that place (or displace) non-human primates within the margins of nature and society. Correspondingly, the incorporation of social science research methods and theory is becoming common practice in primate research and conservation. I have discussed the value of obtaining *ethnographic insight* and the concomitant ethical considerations. These developments are best encapsulated by the now mature sub-field of ethnoprimatology. In a 2014 publication, I, along with a diverse array of leaders in this sub-field, echoed Mark's (2009: 279) observation that "the study of nature is powerful, and power is cultural".[20] Collectively we wrote:

> As anthropologists we are increasingly confronted with complex realities that require diverse and sincerely integrated toolkits, a cultural shift in the loci of power for many of us. Our methodological and theoretical

underpinnings must be drawn from across a range of disciplines and be open to various reflexive critiques and forward looking experiments.

(Malone et al. 2014: 21)

I would argue that the expanding influence of ethnoprimatology within larger academic communities (e.g., primatology and biological anthropology; anthropology in general) is a testament to the hard work and willingness of practitioners such as Agustín Fuentes and Erin Riley to extend themselves as scholars. Some years later, a similar sentiment is arising among some social scientists and conservationists. Striving for meaningful engagement, rather than merely "critique" and "co-optation", Chua et al. (2020) take a progressive look (through the lens of orangutan conservation) at the challenges and possibilities of the conservation–social science relationship. In order to move beyond simplistic incorporation (of social science methods by conservationists) or adversarial critique (of conservation by social scientists), Chua et al. (2020: 54) insist that both parties must:

> remain open to having their most basic assumptions and methods challenged and reworked. This cannot be achieved simply by working better with each other; rather, we also need to work on our own knowledges, practices and relations in potentially destabilising ways. Such a commitment is inherently political in that it entails challenging established epistemological and moral edifices, as well as existing hierarchies and barriers.

Engaging in constructive dialogue, and bringing about disciplinary transformation, is incontrovertibly difficult work. It is also what will be required to stem the tide of global extinction trends. Beyond academic disciplines, there are many fronts where transformative partnerships can be forged and battles won. Here is where dialectical primatologists can play a major role.

Combining dialectical principles with contributions from ethnoprimatology informs the ethical and disciplinary deliberations addressed herein. A dialectical approach does not reduce the relationship with nonhuman primates to a purely utilitarian one. A dialectical approach does integrate a diversity of knowledge by recognising that the natural and the social sciences are inseparable (Levins & Lewontin 1985). This lens allows us to understand that our relations to nonhuman animals are multidimensional, and that while our current practices and methods of engagement with apes produce research results, conservation plans and educational opportunities, they can also harbour a variety of economic, environmental and social problems (Longo & Malone 2006). As evidenced in the various examples above, and indeed in the previous chapters, specific interactions between species are crucial elements in determining the more general trajectories of interspecies relationships. And in these examples we see the respective agencies of humans and animals as serving to either create or collapse dichotomies (e.g., between nature and society; between us and them). Up until this point, I have covered the past and the present of human–ape relationships.

In the next and final chapter, I will extend upon these discussions to envision and describe the future possibilities for life in the hominoid niche.

Notes

1 On 28 May 2016, the day of the incident, the Cincinnati Zoo & Botanical Garden released a media statement which reads, in part: "The Zoo's Dangerous Animal Response Team responded to the life-threatening situation and made the difficult decision to dispatch the gorilla (Harambe)". See http://cincinnatizoo.org/news-releases/media-update-gorilla-world/ for the complete statement.

2 In addition to the behavioural displays related to the heightened intensity of the unfolding events, aspects of protective instincts can be seen in Harambe's behaviour prior to his death. Interestingly, an incident in 1986 involving an adult male gorilla (Jambo) at the Jersey Zoo (formerly Durrell Wildlife Park) was resolved without injury to the gorilla. Jambo initially displayed protective behaviour towards the fallen, unconscious child before distancing himself from the situation which allowed for retrieval of the child. In 1996 at the Brookfield Zoo in Chicago, a similar situation resulted in the safe return of a child that had fallen into the zoo's Tropic World – Africa exhibit. Binti Jua, a hand-reared female gorilla with whom I briefly came to know during my keeper internship with the zoo in 1997, cradled the unconscious child and positioned herself close to an access door. The child was safely retrieved by the husbandry staff. Did Binti Jua's sex and/or the keepers' preconceptions about the behavioural tendencies of females, her role as a new mother or her relationship to the husbandry staff play a role in the outcome? I posit that a combination of these factors were at play.

3 The very ability of animal models to accurately predict biological results in human clinical trials is questioned (Sharpe 1988; Greek & Greek 2000; Shanks et al. 2009), and is based on an "uncontested underlying commonality in the anatomy and physiology of humans and other animals" (Pavelka 2002: 27). The use of nonhuman animals, however, occurs in ways that hide commonality and suppress notions of compassion for other forms of life. The socialist and animal rights activist Henry S. Salt (1851–1939) challenged an illogical dichotomy between nature and society, as well as between humans and nonhuman animals. Salt identified "the hypocrisy of scientists who 'in theory renounced the old-fashioned idea of a universe created for mankind', yet used a position of moral right, ignoring the close relationship that exists between humans and nonhuman animals, to justify the torture of animals" (Clark & Foster 2000: 469). See also Longo and Malone (2006) and Malone et al. (2010).

4 Here, I am borrowing from Irus Braverman's (2014) elegant critique of long-standing conservation discourse that divides conservation sites and strategies along natural versus captive fault lines. From heavily managed wild reserves (or, "captivity in nature") to the idealised, "natural sites" that zoos invite their patrons to imagine, Braverman increasingly foresees the importance of "gray sites" for conservation theory "rather than the schematic black-and-white understanding so paradigmatic of traditional conservation ideologies" (Braverman 2014: 54).

5 Field primatology has traditionally valued the observation of primates in "naturalistic" settings, which sets up a dichotomy between "naturalistic" and "disturbed" locations; that is, those with little perceived human impact and those undeniably impacted by human agency, respectively. This has resulted in the exclusion of most

human–alloprimate interface zones from serious study in primatology and a mini-misation of the role of the human agent. However, over the last three decades field primatologists have increasingly come to terms with the significant human presence at their field sites, and a growing recognition that human impacts matter even in ostensibly "pristine" sites (Wallis & Lee 1999; Fuentes & Wolfe 2002; Malone et al. 2014).

6　In addition to coining the term "ethnoprimatology", Sponsel's explorations of the human niche in Amazonia revealed how the various facets of human life and non-human primate ecologies are co-constituted (e.g., predator–prey relationships; the salience of primates in traditional ecological knowledge) (Sponsel 1997). Wheatley's (1999), *The Sacred Monkeys of Bali*, represents the first, book-length treatment of what would become a time-honoured theme in ethnoprimatological studies – the human–macaque interface. Humans and macaques (genus *Macaca*) rank first and second, respectively, in the extent of their geographic distributions, and therefore have become deeply entangled in many contexts around the world – from densely populated urban areas to agricultural settings, and from temples to tourist sites. See also Riley (2007), Fuentes (2010) and Gumert et al. (2011).

7　In ethnoprimatology, the deductive science of primatology meets the inductive approach of social anthropology (Sommer 2011), resulting in the inclusion of anthropogenic realities as central components of the lives of other primates and their interfaces with humans (Fuentes 2012). As such, a dialectical tension emerges in the attempt to reconcile the tools (methods) of alternatively quantitative and qualita-tive investigations with their juxtaposed justifications (methodologies) – to reduce complexity to something measurable and to describe complex phenomena in their entirety, respectively (Blommaert & Jie 2010; Madden 2010; Malone et al. 2014).

8　Despite the existence of laws to prevent and punish those that are found in private possession of endangered wildlife, an instruction, in 2001, from the Directorate General of Forest Protection and Nature Conservation to regional field offices in effect exempts from prosecution those owners that freely hand over protected animals (*Instruksi Direktur Jendral Perlindungan Hutan dan Konservasi Alam, No. 762/ DJ-IV/ins/121/2001*) (Nijman 2005; Chapter 4, Note 16). Most confiscations are relatively amicable, and typically proceed without confrontation or the allocation of formal penalties.

9　This type of conservation intervention is fraught with ethical issues, including the tendency to construct primates' lives to be as "natural" as possible, even if naturalism comes at the expense of primates' well-being or survival. In gibbon rehabilitation circles, practitioners debated whether to prioritise the welfare of released gibbons or the "naturalness" of the release-site's condition. For example, the presence of predators such as leopards at release-sites are flexibly described as both "indicators of a suitable release site" and also "an operational liability" given the naiveté of gibbon release candidates. The latter characterisation was then contextualised as being preferable to the presence of human hunters, reflecting the idea that a "nat-ural" death from leopard predation was preferable to death at the ("unnatural") hand of humanity (Palmer & Malone 2018: 840).

10　Journal editor Alan Smart states:

> Their [Malone et al. 2014] willingness to adopt ethnographic methods and incorporate people into their primatological research programs should be an inspiration to the rest of the discipline in finding ways to cooperate effectively

to deal with crucial research challenges that require multiple sets of skills not confined to a single sub discipline.

(Smart 2014: 5)

See also Dore (2017), Remis and Jost Robinson (2017) and Shaffer et al. (2017) for additional examples.

11 Carla Willig (2013: 10) defines two types of "reflexivity":

> Personal reflexivity' involves reflecting upon the ways in which our own values, experiences, interests, beliefs, political commitments, wider aims in life and social identities have shaped the research. It also involves thinking about how the research may have affected and possibly changed us, as people and as researchers. "Epistemological reflexivity" requires us to engage with questions such as: How has the research question defined and limited what can be "found?" How has the design of the study and the method of analysis "constructed" the data and the findings? How could the research question have been investigated differently? To what extent would this have given rise to a different understanding of the phenomenon under investigation? Thus, epistemological reflexivity encourages us to reflect upon the assumptions (about the world, about knowledge) that we have made in the course of the research, and it helps us to think about the implications of such assumptions for the research and its findings.

12 The One Health (OH) perspective sees the interdependence of human, animal and environmental health, and conceptualises human–animal–environment interactions as inherently linked to the emergence and expression of infectious disease (Landford & Nunn 2012; Degeling et al. 2015). By definition, more of an ecological than an anthropocentric approach, OH models health risks, benefits and mitigations in aggregate among species, and between species and their environments.

13 Our response to the target article raises additional, methodological challenges to the feasibility/implementation of such a "field trial". Edwards and colleagues cite methods proposed by Walsh et al. (2017) with respect to oral administration of the vaccine, yet it is unclear how this could be realised by noninvasive means, in field conditions. "Controlling rates of vaccine uptake" (Walsh et al. 2017: 6) strikes us as particularly difficult to achieve, even by the methods proposed (e.g., remote trigger). Walsh and colleagues also describe a monitoring regime that entails frequent blood draws plus body mass measurements. The need for genuinely noninvasive methods is especially acute given the possible role of the stress response in mediating immunity (Walsh et al. 2017). Second, the suitability of an Ebola virus (EBOV) vaccine is problematised by the virus's high rate of non-synonymous mutation. Mutations have been observed in regions of the genome that code for the surface glycoprotein (Gire et al. 2014) of EBOV, which could be particularly problematic for vaccine development. This could significantly curtail the potential benefits of a vaccine for nonhuman primates (Addison & Malone 2018: 54).

14 Alison Wade (AW) conducted research in a 50 km² area of the Mone-Oku Forest complex, Manyu Division, Southwest Cameroon between May 2014 and March 2015. As there is no habituated population of either ape, night nests were used to gain insights into the apes' behaviour and how behaviour is shaped by ecological interactions. Specifically, the area is an unclassified forest corridor located between Mone Forest Reserve and the Mbulu Forest that provides tenuous links

in the north to Takamanda National Park and the Kagwene Gorilla Sanctuary. The forest is topographically diverse as elevations range between 130 metres and 1000 metres and vegetation falls into the Guinea Congolian type (White 1983). It is a mosaic landscape of villages, farmland, secondary and primary forest. AW carried out 50 semi-structured interviews encompassing a range of topics including forest use, farming practices, economic pathways and village history. Interview data were supplemented with participant observation focusing on farming and forest use, providing further insight into which resources people use and how they use them (Riley & Ellwanger 2013; Malone et al. 2017). For more complete reporting of AW's methods and results, see Wade et al. (2019) and Wade (2020).

15 The Wildlife Conservation Society, through its Gorilla Guardian Programme (GGP), has been consistently monitoring both ape species in the region since 2009. The GGP promotes the conservation of both ape species in villages outside of protected areas and provides individuals nominated within each village with tracking and ecological training (Jameson 2012; Wade et al. 2019: 55).

16 In Cameroon, Forest, Wildlife and Fisheries Regulations (Law #94/01) identify both the Cross River gorilla and the Nigeria-Cameroon chimpanzees as Category A species. Under Section 78(2) Category A species are fully protected against hunting, capture or sale: in whole or in part. However, Sections 82 and 83 do provide exceptions to this if the animals become a threat to people or property. Section 82 reads "In cases where animals constitute a danger or cause damage to persons and/or property, the service in charge of wildlife may undertake to hunt them down under conditions laid down by order of the minister in charge of wildlife". Section 83(1) concerns the protection of property and reads "No persons may be charged with breach of hunting regulations as concerns protected animals if his act was dictated by the urgent need to defend himself, his livestock or crops".

17 The New Zealand Animal Welfare Act of 1999 states that no person can carry out "research, testing, or teaching involving the use of a nonhuman hominid" unless it has been determined by the director-general of the Ministry for Primary Industries to be either: (a) in the best interests of the individual nonhuman hominid; or (b) in the best interests of the species to which the individual nonhuman hominid belongs, provided that the benefits to the species are not outweighed by likely harm to the individual (Parliamentary Counsel Office 1999).

18 In this same vein, authors have drawn on Foucault (1977) to consider the ways in which the zoo's structure and layout act in a similar way to prisons to reinforce divisions between keeper and kept, watcher and watched (e.g., Mullan & Marvin 1987; Malamud 1998; Beardsworth & Bryman 2001; Acampora 2005).

19 See Bron Taylor's contribution "The value of a gorilla vs. a human", available at: www.huffingtonpost.com/bron-taylor/the-value-of-a-gorilla_b_10214928.html.

20 The more complete quote from Marks (2009: 279): "The study of nature is powerful, and power is cultural. Engaging cultural issues is essential for understanding science; it is not the antithesis of science, as spirituality is".

References

Acampora R. 2005. Zoos and eyes: Contesting captivity and seeking successor practices. *Society and Animals* 13(1): 69–88.

Addison C, Malone N. 2018. An experimental ethics, but an ethical experiment? Anthropological perspectives on using unproven vaccines on endangered primates. *The American Journal of Bioethics* 18(10): 53–55.

Asquith PJ. 2011. Of bonds and boundaries: What is the modern role of anthropomorphism in primatological studies? *American Journal of Primatology* 73(3): 238–244.

Baker LR. 2002. Guidelines for nonhuman primate re-introductions. *Newsletter of the Re-introduction Specialist Group of the IUCN's Species Survival Commission (SSC)* 21: 1–32.

Beardsworth A, Bryman A. 2001. The wild animal in late modernity. *Tourist Studies* 1(1): 83–104.

Beck BB, Rodrigues M, Unwin S, Travis D, Stoinski T. 2007. *Best Practice Guidelines for the Re-Introduction of Great Apes* (Occasional Paper of the IUCN Species Survival Commission No. 35). Gland: SSC Primate Specialist Group of the World Conservation Union.

Behie A, Teichroeb J, Malone N. (Eds.). 2019. *Primate Research and Conservation in the Anthropocene* (Vol. 82). Cambridge: Cambridge University Press.

Berger J. 1980. *About Looking.* New York, NY: Pantheon Books.

Bergl RA, Warren Y, Nicholas A, Dunn A, Imong I, Sunderland-Groves JL, Oates JF. 2012. Remote sensing analysis reveals habitat, dispersal corridors and expanded distribution for the critically endangered Cross River gorilla *Gorilla gorilla diehli. Oryx – The International Journal of Conservation* 46: 278–289.

Bernard HR. 2011. *Research Methods Anthropology: Qualitative and Quantitative Methods,* 5th ed. Maryland, MD: AltaMira Press.

Blommaert J, Jie D. 2010. *Ethnographic Fieldwork: A Beginner's Guide.* Bristol: Multilingual Matters.

Braverman I. 2014. Conservation without nature: The trouble with in situ versus ex situ conservation. *Geoforum* 51: 47–57.

Bulbeck C. 2005. *Facing the Wild: Ecotourism, Conservation and Animal Encounters.* London: Earthscan.

Bullo K. 2015. *Reaching for the Canopy: A Zoo-Born Orangutan's Journey into the Wild.* Crawley: UWA Publishing.

Chua L, Harrison ME, Fair H, Milne S, Palmer A, Rubis J, Thung P, Wich S, Büscher B, Cheyne SM, Puri RK. 2020. Conservation and the social sciences: Beyond critique and co-optation. A case study from orangutan conservation. *People and Nature* 2(1): 42–60.

Clark B, Foster JB. 2000. Henry S. Salt, socialist, animal rights activist: An introduction to Salt's *A Lover of Animals. Organization and Environment* 13: 468–473.

Cocks L, Bullo K. 2008. The processes for releasing a zoo-bred Sumatran orang-utan *Pongo abelii* at Bukit Tigapuluh National Park, Jambi, Sumatra. *International Zoo Yearbook* 42(1): 183–189.

Corbey R, Theunissen B. 1995. *Ape, Man, Apeman: Changing Views since 1600.* Evaluative proceedings of the symposium Ape, man, apeman: Changing views since 1600, Leiden, the Netherlands, June 28–July 1, 1993. Leiden: Department of Prehistory, Leiden University.

Courchamp F, Angulo E, Rivalan P, Hall RJ, Signoret L, Bull L, Meinard Y. 2006. Rarity value and species extinction: The anthropogenic Allee effect. *PLoS Biology* 4(12): e415.

Degeling C, Johnson J, Kerridge I, Wilson A, Ward M, Stewart C, Gilbert G. 2015. Implementing a One Health approach to emerging infectious disease: Reflections on the socio-political, ethical and legal dimensions. *BMC Public Health* 15(1): 1307.

Dore KM. 2017. Navigating the methodological landscape: Ethnographic data expose the nuances of the Monkey Problem in St. Kitts, West Indies. In Dore KM, Riley EP, Fuentes A. (Eds.), *Ethnoprimatology: A Practical Guide to Research at* the Human–Nonhuman Primate Interface (pp. 219–231). Cambridge: Cambridge University Press.

Dore KM, Riley EP, Fuentes A. (Eds.). 2017. *Ethnoprimatology: A Practical Guide to Research at the Human-Nonhuman Primate Interface* (Cambridge Studies in Biological and Evolutionary Anthropology). Cambridge: Cambridge University Press.

Dunn A, Bergl R, Byler D, Eben-Ebai S, Etiendem DN, Fotso R, Ikfuingei R, Imong I, Jameson C, Macfie L, Morgan B, Nchanji A, Nicholas A, Nkembi L, Omeni F, Oates J, Pokempner A, Sawyer S, Williamson EA. 2014. *Revised Regional Action Plan for the Conservation of the Cross River Gorilla (Gorilla gorilla diehli): 2014–2019*. New York, NY: IUCN/SSC Primate Specialist Group and Wildlife Conservation Society.

Edwards SJL, Norell CH, Illari P, Clarke B, Neuhaus CP. 2018. A radical approach to Ebola: Saving humans and other animals. *The American Journal of Bioethics* 18(10): 35–42.

Ellwanger AL, Riley EP, Niu K, Tan CL. 2017. Using a mixed-methods approach to elucidate the conservation implications of the human-primate interface in Fanjingshan National Nature Reserve, China. In Dore KM, Riley EP, Fuentes A. (Eds.), *Ethnoprimatology: A Practical Guide to Research at the Human–Nonhuman Primate Interface* (pp. 257–270). Cambridge: Cambridge University Press.

Fedigan LM. 2010. Ethical issues faced by field primatologists: Asking the relevant questions. *American Journal of Primatology* 72(9): 754–771.

FORINA. 2013. *Optimising Orangutan Reintroduction: Report of a Workshop*. Bogor: Workshop organised by FORINA, supported by Arcus Foundation.

Foucault M. 1977. *Discipline and Punish: The Birth of the Prison*. New York, NY: Pantheon Books.

Fuentes A. 2010. Naturalcultural encounters in Bali: Monkeys, temples, tourists, and ethnoprimatology. *Cultural Anthropology* 25: 600–624.

Fuentes A. 2012. Ethnoprimatology and the anthropology of the human-primate interface. *Annual Review of Anthropology* 41: 101–117.

Fuentes A, Wolfe LD. 2002. *Primates Face to Face: The Conservation Implications of Human-Nonhuman Primate Interconnections*. Cambridge: Cambridge University Press.

Gezon LL. 2006. *Global Visions, Local Landscapes: A Political Ecology of Conservation, Conflict, and Control in Northern Madagascar*. Lanham, MD: Altamira Press.

Gire SK, Goba A, Andersen KG, Sealfon RS, Park DJ, Kanneh L, Jalloh S, Momoh M, Fullah M, Dudas G, Wohl S. 2014. Genomic surveillance elucidates Ebola virus origin and transmission during the outbreak. *Science* 345(6202): 1369–1372.

Greek CR, Greek JS. 2000. *Sacred Cows and Golden Geese*. New York, NY: Continuum Publishing Group.

Gumert MD, Fuentes A, Jones-Engel L. (Eds.). 2011. *Monkeys on the Edge: Ecology and Management of Long-tailed Macaques and their Interface with Humans* (Vol. 60). Cambridge: Cambridge University Press.

Hanson KT, Riley EP. 2018. Beyond neutrality: The human–primate interface during the habituation process. *International Journal of Primatology* 39(5): 852–877.

Haraway D. 2008. *When Species Meet*. Minneapolis, MN: University of Minnesota Press.

Hosey GR. 2005. How does the zoo environment affect the behaviour of captive primates? *Applied Animal Behaviour Science* 90(2): 107–129.

Ingold T. 2012. Toward an ecology of materials. *Annual Review of Anthropology* 41: 427–442.

Isaac NJ, Cowlishaw G. 2004. How species respond to multiple extinction threats. *Proceedings of the Royal Society of London. Series B: Biological Sciences* 271(1544): 1135–1141.

Jacobson SK. 2010. Effective primate conservation education: Gaps and opportunities. *American Journal of Primatology* 72: 414–419.

Jameson C. 2012. Gorilla Guardian update: Expansion of the community-based monitoring network. *Gorilla Journal* 45: 13–15.

Jones RC, Greek R. 2014. A review of the Institute of Medicine's analysis of using chimpanzees in biomedical research. *Science and Engineering Ethics* 20(2): 481–504.

Jost Robinson CA. 2012. *Beyond Hunters and Hunted: An Integrative Anthropology of Human-Wildlife Dynamics and Resource Use in a Central African Forest.* PhD Thesis, Purdue University, Indiana, USA.

King T, Courage A. 2008. Western gorilla re-introduction to the Batéké Plateau region of Congo and Gabon. In *Global Re-Introduction Perspectives: Re-Introduction Case-Studies From Around the Globe* (pp. 217–220). Abu Dhabi: IUCN/SSC Re-introduction Specialist Group.

Kirksey SE, Helmreich S. 2010. The emergence of multispecies ethnography. *Cultural Anthropology* 25: 545–576.

Kleiman DG, Beck BB, Dietz JM, Dietz LA. 1991. Costs of a re-introduction and criteria for success: Accounting and accountability in the Golden Lion Tamarin Conservation Program. *Symposia of the Zoological Society of London* 62:125–142.

Knight J. 2005. Feeding Mr. Monkey: Cross-species food "exchange" in Japanese monkey parks. In Knight J. (Ed.), *Animals in Person: Cultural Perspectives on Animal-Human Intimacy* (pp. 231–254). Oxford: Berg.

Knight J. 2017. Introduction to Part II – Following the data: Incorporating ethnography. In Dore KM, Riley EP, Fuentes A. (Eds.), *Ethnoprimatology: A Practical Guide to Research at the Human-Nonhuman Primate Interface* (pp. 171–175). Cambridge Studies in Biological and Evolutionary Anthropology. Cambridge: Cambridge University Press.

Kopnina H. 2017. Beyond multispecies ethnography: Engaging with violence and animal rights in anthropology. *Critique of Anthropology* 37(3): 333–357.

Landford J, Nunn M. 2012. Good governance in "One Health" approaches. *Revue Scientifique et Technique (International Office of Epizootics)* 31(2): 561–575.

Lee P. 2010. Sharing space: Can ethnoprimatology contribute to the survival of non-human primates in human-dominated globalized landscapes? *American Journal of Primatology* 72(10): 925–931.

Levins R, Lewontin R. 1985. *The Dialectical Biologist.* Cambridge, MA: Harvard University Press.

Longo SB, Malone N. 2006. Meat, medicine, and materialism: A dialectical analysis of human relationships to nonhuman animals and nature. *Human Ecology Review* 13: 111–121.

MacClancy J. 2013. Covering our backs, or covering all bases? An ethnography of URECs. In *Ethics in the Field: Contemporary Challenges* (pp. 175–190). New York, NY: Berghahn Books.

MacKinnon KC, Riley EP. 2010. Field primatology of today: Current ethical issues. *American Journal of Primatology* 72(9): 749–753.

MacKinnon KC, Riley EP. 2013. Contemporary ethical issues in field primatology. In *Ethics in the Field: Contemporary Challenges* (pp. 98–107). New York, NY: Berghahn Books.

Madden R. 2010. *Being Ethnographic: A Guide to the Theory and Practice of Ethnography*. London: Sage.

Malamud R. 1998. *Reading Zoos: Representations of Animals and Captivity*. New York, NY: New York University Press.

Malone N, Fuentes A, Purnama AR, Adi Putra IM W. 2004. Displaced hylobatids: Biological, cultural, and economic aspects of the primate trade in Java and Bali, Indonesia. *Tropical Biodiversity* 40(1): 41–49.

Malone NM, Fuentes A, White FJ. 2010. Ethics commentary: Subjects of knowledge and control in field primatology. *American Journal of Primatology* 72(9): 779–784.

Malone N, Palmer A. 2014. Ethical issues within human–alloprimate interactive zones. In *Engaging with Animals: A Shared Existence* (pp. 21–38). Sydney: University of Sydney Press.

Malone N, Palmer A, Wade AH. 2017. Incorporating the ethnographic perspective: The value, process, and responsibility of working with human participants. In Dore KM, Riley EP, Fuentes A. (Eds.), *Ethnoprimatology: A Practical Guide to Research at the Human–Nonhuman Primate Interface* (pp. 176–189). Cambridge: Cambridge University Press.

Malone N, Wade AH, Fuentes A, Riley EP, Remis MJ, Jost Robinson C. 2014. Ethnoprimatology: Critical interdisciplinarity and multispecies approaches in anthropology. *Critique of Anthropology* 38: 8–29.

Marks J. 2009. *Why I Am Not a Scientist: Anthropology and Modern Knowledge*. Berkeley: University of California Press.

McKinney T. 2015. A classification system for describing anthropogenic influence on nonhuman primate populations. *American Journal of Primatology* 77(7): 715–726.

McLennan MR, Hill CM. 2013. Ethical issues in the study of conservation of an African great ape in an unprotected, human–dominated landscape in Western Uganda. In *Ethics in the Field: Contemporary Challenges* (pp. 42–66). New York, NY: Berghahn Books.

Morgan BJ, Adeleke A, Bassey T, Bergl R, Dunn A, Fotso R, Gadsby E, Gonder MK, Greengrass E, Koulagna DK, Mbah G. 2011. *Regional Action Plan for the Conservation of the Nigeria-Cameroon Chimpanzee (Pan troglodytes ellioti)*. San Diego, CA: IUCN/SSC Primate Specialist Group and Zoological Society of San Diego.

Mullan B, Marvin G. 1987. *Zoo Culture*. London: Weidenfeld & Nicolson.

Nading AM. 2014. *Mosquito Trails: Ecology, Health, and the Politics of Entanglement*. Berkeley: University of California Press.

Nijman V. 2005. *In Full Swing. An Assessment of the Trade in Gibbons and Orang-utans on Java and Bali, Indonesia*. Kuala Lumpur: TRAFFIC Southeast Asia.

Oates JF. 1999. *Myth and Reality in the Rain Forest: How Conservation Strategies are Failing in West Africa*. Berkeley: University of California Press.

Ogden LA, Hall B, Tanita K. 2013. Animals, plants, people, and things: A review of multispecies ethnography. *Environment and Society* 4(1): 5–24.

Palmer A. 2012. *Keeper/Orangutan Interactions at Auckland Zoo: Communication, Friendship, and Ethics Between Species*. MA Thesis, University of Auckland.

Palmer A. 2020. *Ethical Debates in Orangutan Conservation*. London: Routledge.

Palmer A, Malone N. 2018. Extending ethnoprimatology: Human–alloprimate relationships in managed settings. *International Journal of Primatology* 39(5): 831–851.

Palmer A, Malone N, Park J. 2015. Accessing orangutans' perspectives: Interdisciplinary methods at the human/animal interface. *Current Anthropology* 56(4): 571–578.

Palmer A, Malone N, Park J. 2016. Caregiver/orangutan relationships at Auckland Zoo: Empathy, friendship, and ethics between species. *Society & Animals* 24(3): 230–249.

Parliamentary Counsel Office. 1999. Animal Welfare Act 1999 no. 142, reprint as at 7 July 2010. [Online] Available: www.legislation.govt.nz/act/public/1999/0142/latest/DLM51206.html.

Pavelka MV. 2002. Resistance to the cross-species perspective in anthropology. In Fuentes A, Wolf LD. (Eds.), *Primates Face to Face: The Conservation Implications of Human-Nonhuman Primate Interconnections* (pp. 25–44). Cambridge: Cambridge University Press.

Remis MJ, Hardin R. 2009. Transvalued species in an African forest. *Conservation Biology* 23(6): 1588–1596.

Remis MJ, Jost Robinson CA. 2012. Reductions in primate abundance and diversity in a multi-use protected area: Synergistic impacts of hunting and logging in a Congo Basin forest. *American Journal of Primatology* 74: 602–612.

Remis MJ, Jost Robinson CA. 2017. Nonhuman primates and "others" in the Dzanga Sangha reserve: The role of anthropology and multispecies approaches in ethnoprimatology. In Dore KM, Riley EP, Fuentes A. (Eds.), *Ethnoprimatology: A Practical Guide to Research at the Human–Nonhuman Primate Interface* (pp. 190–205). Cambridge: Cambridge University Press.

Rijksen HD, Meijaard E. 1999. Rehabilitation of orang-utans. In *Our Vanishing Relative: The Status of Wild Orang-Utans at the Close of the Twentieth Century* (pp. 151–175). Dordrecht: Kluwer Academic Publishers.

Riley EP. 2006. Ethnoprimatology: Toward reconciliation of biological and cultural anthropology. *Ecological and Environmental Anthropology* 2(2): 75–86.

Riley EP. 2007. The human–macaque interface: Conservation implications of current and future overlap and conflict in Lore Lindu National Park, Sulawesi, Indonesia. *American Anthropologist* 109(3): 473–484.

Riley EP. 2010. The importance of human-macaque folklore for conservation in Lore Lindu National Park, Sulawesi, Indonesia. *Oryx* 44(2): 235–240.

Riley EP. 2013. Contemporary primatology in anthropology: Beyond the epistemological abyss. *American Anthropologist* 115(3): 411–422.

Riley EP. 2018. The maturation of ethnoprimatology: Theoretical and methodological pluralism. *International Journal of Primatology* 39(5): 705–729.

Riley EP. 2020. *The Promise of Contemporary Primatology.* New York, NY: Routledge.

Riley EP, Bezanson M. 2018. Ethics of primate fieldwork: Toward an ethically engaged primatology. *Annual Review of Anthropology* 47: 493–512.

Riley EP, Ellwanger A. 2013. Methods in ethnoprimatology: Exploring the human-nonhuman primate interface. In *Primate Ecology and Conservation: A Handbook of Techniques* (pp. 128–150). Oxford: Oxford University Press.

Riley EP, Priston NEC. 2010. Macaques in farms and folklore: Exploring the human–nonhuman primate interface in Sulawesi, Indonesia. *American Journal of Primatology* 72(10): 848–854.

Rosen N, Buyers O. 2002. *Orangutan Conservation and Reintroduction Workshop: Final Report.* Workshop held 19–22 June 2002, Palangkaraya, Indonesia. Apple Valley, MN: IUCN/SSC Conservation Breeding Specialist Group.

Russell WMS, Burch RL. 1959. *The Principles of Humane Experimental Technique.* London: Methuen & Co.

Russon AE. 2009. Orangutan rehabilitation and reintroduction. In Wich S, Utami Atmoko S, Setia TM, van Schaik C. (Eds.), *Orangutans: Geographic Variation in Behavioral Ecology and Conservation* (pp. 327–350). Oxford: Oxford University Press.

Shaffer CA, Marawanaru E, Yukuma C. 2017. An ethnoprimatological approach to assessing the sustainability of nonhuman primate subsistence hunting of indigenous Waiwai in the Konashen Coummunity Owned Conservation Area, Guyana. In Dore KM, Riley EP, Fuentes A. (Eds.), *Ethnoprimatology: A Practical Guide to Research at the Human–Nonhuman Primate Interface* (pp. 232–250). Cambridge: Cambridge University Press.

Shanks N, Greek R, Greek J. 2009. Are animal models predictive for humans? *Philosophy, Ethics, and Humanities in Medicine* 4(1): 2.

Sharpe R. 1988. *The Cruel Deception: The Use of Animals in Medical Research.* Wellingborough: Thorsons Publishing Group.

Siriwat P, Nekaris KAI, Nijman V. 2019. The role of the anthropogenic Allee effect in the exotic pet trade on Facebook in Thailand. *Journal for Nature Conservation* 51: 125726.

Smart A. 2014. Critical perspectives on multispecies ethnography. *Critique of Anthropology* 34(1): 3–7.

Smits W, Heriyanto, Ramono W. 1995. A new method for rehabilitation of orangutans in Indonesia: A first overview. In Galdikas BM, Nadler RD, Rosen N, Sheeran L. (Eds.), *The Neglected Ape* (pp. 69–78). New York, NY: Springer.

Solomon DA. 2016. Interpellation and affect: Activating political potentials across primate species at Jakhoo Mandir, Shimla. *Humanimalia* 8(1): 1–34.

Sommer V. 2011. The anthropologist as a primatologist: Mental journeys of a fieldworker. In MacClancy J, Fuentes A. (Eds.), *Centralizing Fieldwork: Critical Perspectives from Primatology, Biological and Social Anthropology* (pp. 32–48). New York, NY; Oxford: Berghahn Books.

Sponsel LE. 1997. The human niche in Amazonia: Explorations in ethnoprimatology. In Kinzey WG. (Ed.), *New World Primates* (pp. 143–165). New York, NY: Aldine De Gruyter.

Tongco MDC. 2007. Purposive sampling as a tool for informant selection. *Ethnobotany Research & Applications* 5: 147–158.

Tyson E. 1699. *Orang-outang, sive Homo Sylvestris: Or, The Anatomy of a Pygmie Compared with that of a Monkey, an Ape and a Man.* London: Bennet.

Wade AH. 2012. *Exploring Ethnoprimatology.* MA Portfolio, University of Auckland, New Zealand.

Wade AH. 2020. *Shared Landscapes: The Human-Ape Interface within the Mone-Oku Forest, Cameroon.* PhD Thesis, University of Auckland, New Zealand.

Wade A, Malone N, Littleton J, Floyd B. 2019. Uneasy neighbours: Local perceptions of the Cross River gorilla and Nigeria-Cameroon chimpanzee in Cameroon. In Behie A, Teichroeb J, Malone N. (Eds.), *Primate Research and Conservation in the Anthropocene* (Vol. 82, pp. 52–73). Cambridge: Cambridge University Press.

Waller MT (Ed.). 2016. *Ethnoprimatology: Primate Conservation in the 21st Century.* New York, NY: Springer.

Wallis J, Lee DR. 1999. Primate conservation: The prevention of disease transmission. *International Journal of Primatology* 20: 803–826.

Walsh PD, Kurup D, Hasselschwert DL, Wirblich C, Goetzmann JE, Schnell MJ. 2017. The final (oral Ebola) vaccine trial on captive chimpanzees? *Scientific Reports* 7: 43339.

Wedana Adi Putra IM, Jeffery S. 2016. Reinforcing the Javan silvery gibbon population in the Mount Tilu Nature Reserve, West Java, Indonesia. Programme of the XXVIth Congress of the International Primatological Society, Chicago, IL, USA, August 21–27, 2016.

White F. 1983. *The Vegetation of Africa, a Descriptive Memoir to Accompany the UNESCO/AETFAT/UNSO Vegetation Map of Africa (3 Plates, Northwestern African, Northeastern Africa, and Southern Africa, 1:5,000,000).* Paris: UNESCO.

Willig C. 2013. *Introducing Qualitative Research in Psychology*, 3rd ed. Berkshire: Open University Press.

6 Conclusion

The future of life in the hominoid niche

> To predict the threshold beyond which ape populations are unable to accommodate human presence and activities, and local people can no longer tolerate apes and other wildlife, research is needed on populations at different stages of the anthropogenic continuum. To do this, we should abandon a simplistic "anthropogenic-or-not" approach and instead identify variables, including human activities and customs, which accurately characterise the different types of anthropogenic landscape, and determine their influence on the behaviour of apes and other wildlife.
>
> (Kimberley Hockings et al. 2015: 220)

Make no mistake, the living apes are in serious trouble. The large brained apes with protracted life histories, once dominant in primate communities of the past, now face a sustained barrage of threats from the most "successful" hominoid that has ever lived – ourselves. Apes possess an ability to adjust their behaviour to minor shifts in ecological and demographic conditions via social innovation and learning (van Schaik 2013; Hockings et al. 2015). However, these same species are also slow-reproducing and are therefore vulnerable to rapid environmental alterations, such as those brought about by humans. Arguably, the ability to completely contextualise our own existence depends upon their survival. Our fates are inextricably linked. Do we truly understand the complexities of sharing the planet with our closet living biological relatives? Have we overestimated (or underestimated) the behavioural and ecological flexibility of our endangered ape kin? The living hominoids share an evolved emphasis on: (a) cognition and learning, underpinned by a particular neurological complexity, especially in the cortical areas of the brain; (b) extended life history phases, such as long periods of sub-adulthood; and (c) the formation of complex social relationships. Over time, humankind's ratcheting-up from this evolutionary baseline has resulted in the dramatic shaping of ecological systems to suit our needs. But it has come at a price. Is our "success" predicated upon the systematic dismantling of the very conditions where such evolutionary trajectories were forged (and where our ape kin still reside)? Our ability to address these questions will determine the fate of the planet's remaining apes.

DOI: 10.4324/9780367211349-6

In the anthropocene, humanity plays a central role in systemically shaping planetary processes. As Ritchie and Knight (2016: 213) state, the "anthropocene" can be debated along formal stratigraphic lines concerned with "the recognition, characterisation and onset of the anthropocene as a geological epoch", and also conceptually "to recognise and discuss global environmental change, including its historical roots and philosophical implications". My scholarly engagement with the "anthropocene" is decidedly focused upon the latter – the more conceptual debate – especially as it pertains to local manifestations of the dynamic forces of environmental change.[1] The order Primates is in peril as the ongoing, human-caused "sixth mass extinction" pushes the planet's vertebrates to the brink (Estrada et al. 2017; Ceballos et al. 2020).[2] Of particular concern given the present topic (the future of hominoid species), is that despite the dedicated efforts of numerous scientists and conservationists, we will likely bear witness to the extinction of several ape species in the coming decades. Estrada et al. (2017) identified a distinct phylogenetic signal within primate clades (including apes) indicating the compounding effect of intrinsic risk factors and extrinsic, anthropogenic threats. For apes, where birth rates barely exceed death rates, the capacity for population growth and rebound is severely restricted (Wich et al. 2004; Littleton 2005; Bronikowski et al. 2016; Kramer 2019). With agricultural intensification on the rise, forested habitats diminishing, and a host of human–nature conflicts to mitigate (Laurance et al. 2014; Estrada et al. 2017), the present generation of primatologists must immediately come to terms with the consequences of our actions (and inactions). How will interested scientists and conservation activists unite to stem the tide of humanity's impact? How can we summon the collective political strength to leverage governments and corporations to mount comprehensive responses to the extinction crisis before it is too late? In search of answers to these challenging questions, I will briefly revisit themes from the previous chapters, with a focused attention on identifying both internal contradictions and unifying themes. Then, I construct a synthesis aimed towards creating the conditions for hominoid coexistence, and suggest some tangible future directions.

Recapitulation

Dialectics and influences

At the beginning of the book I laid out the influences and principles that have guided my thinking as a self-described, dialectical primatologist. With insights arising from my particular constellation of research engagements with captive, displaced and wild apes, I have sought to articulate new perspectives on primate (specifically hominoid) evolution, behaviour and conservation. These perspectives are grounded in anthropological, philosophical and sociological traditions, and explicitly embrace the critiques of reductionist biology advanced so powerfully by the likes of Stephen Jay Gould, Richard Levins and Richard Lewontin. Opportunities to further my studies and research, and my theoretical baselines are attributed to the generous mentoring of Agustín Fuentes, whom I have known for over 20 years. Additionally, the time I spent at the University of Oregon (during my doctoral studies) afforded me opportunities to engage in conversations with the editors of, and frequent contributors to, the *Monthly Review* – the long-established, independent socialist periodical.[3] Conversations with the likes of John Bellamy Foster, Brett Clark, Joseph Fracchia, Stefano Longo and Richard York added value to my training as an anthropologist in the "four-field" tradition. My critical lens therefore, comprising equal parts anthropological holism, evolutionary biology and critique of capitalism, has led me to view certain modes of production within primatological practice, and the production of scientific knowledge more generally, in the dialectical tradition.[4]

The dialectical principles I established in Chapter 1, specifically tailored to the present topic, were glossed generally as *emergence, constrained totipotentiality* and *constructed compatibilities*. In the research examples discussed throughout the book, I demonstrate how attention to the dynamics among various levels of organisation, the channelling of variation along general pathways (as opposed to intensely selected deviations) and the analytical goal of achieving synthesis, can lead to new perspectives on ape social systems, behavioural flexibility and underlying evolutionary mechanisms. For ape species, with a high degree of intraspecific behavioural variation, *constrained totipotentiality* refers to how this variation plays out over both evolutionary and ecological timescales. Additionally, the same mode of thinking can produce critical conservation insights, especially with respect to the ways in which we conceptualise our relationships to our fellow hominoids. Such conceptualisations, I argue, can impact the types of questions we ask and the interpretive lenses through which we analyse our data; and they can determine the conservation interventions we propose and implement. Conceptualisations of humankind's relationship to nonhuman animals are often imbued with notions of domination and mastery, or indeed present humans in a proprietary role to animals and the whole of nature. Within this view of nature, the exploitation of the natural world is acceptable (Longo & Malone 2006). In contrast, a dialectical approach to the natural world sees humanity as enveloped in natural processes, not outside of them. Nature is a dialectical process and "opposing forces lie at the base of the evolving physical and biological world" (Levins & Lewontin 1985: 280). Organisms are both subjects and objects, causes and effects of their environment. There is an interactive effect occurring that influences all aspects of the environment including, of course, humans. A materialist, dialectical and co-evolutionary perspective is necessary for understanding the human society/nature relationship (Foster 2000).

From the Miocene to the margins, again

In Chapter 2, I covered approximately 30 million years of primate evolutionary history, with special attention to the adaptive capacities and constraints of the hominoids. I set out to establish a baseline for the book by providing clear answers to the following questions: who are the apes and how are their evolutionary trajectories distinct from the other primates? What was the impetus for the evolution of the basic, ape "life-way", and how have humans extended this trajectory? What are the present threats to ape populations, and what are the limits of ape resilience in the face of humankind's expanding reach? A warming climatic trend at the Oligocene/Miocene transition, coupled with an increasing distribution of heterogeneous forest and woodland habitats, provided a suitable context for emerging radiations of, initially medium-sized, arboreal catarrhine (paleotropical anthropoid) primates: the Hominoidea (apes) and the Cercopithecoidea (monkeys). These relatively aseasonal forests provided a consistent supply of readily available energy in the form of ripe fruit. For the emerging ape clade, this stability paved the way for an adaptive life-way that included protracted life histories and post-cranial specialisations (e.g., arboreal bridging and suspension; forelimb mobility and propulsion) that ensured efficient access to the high-quality diets on offer (Temerin & Cant 1983; Kelly 1997; Jablonski 2005). Ultimately, the varying evolutionary trajectories of paleotropical monkeys and apes, under ecologically similar conditions, established a terminal-Miocene baseline for the fates of these lineages in the Pliocene, Pleistocene, Holocene and beyond. The Miocene belonged to the apes; the more recent past generally favours, in terms of both species richness and abundance, the resilient cercopithecoids. That being said, the present – the anthropocene – poses significant risk to even the most successful of nonhuman primates.

This chapter concludes by situating contemporary humans among the evolutionary backdrop of hominoid evolution. Marks (2012: 100) succinctly states that: "our evolutionary relationship to the apes could probably be more usefully seen as one in which continuity and discontinuity coexist in tension with one another; and whose cultural meanings suffuse the data produced on their behalf". By definition, ape research and conservation (as fellow hominoids) requires us to consider our subjectivities and biases. Humans share with apes an evolved capacity to meet ecological challenges with social solutions. In this way, humankind's extensive ability to shape and construct ecological worlds arises from a common "ape" baseline. Humans, however, through a process of biocultural evolution, have dramatically altered the trajectory of our species' global impact by way of both enhanced reproductive productivity and technological sophistication. Therefore, how we choose to view ourselves in relation to the apes lies at the heart of our broader articulation to the whole of nature.[5] Will our knowledge of extinct and extant hominoids reify notions of human domination and mastery, or will we control the use of our species-specific attributes in order to engineer space and time for all extant hominoids to continue their respective evolutionary courses?

Emergence and theory in context

I began Chapter 3 by acknowledging that contemporary landscapes do not exist independent of humans, and human societies do not exist independent of a biophysical context (Fuentes & Baynes-Rock 2017;York & Longo 2017). Still, many aspects of nonhuman hominoid "ecologies" emerged prior to (i.e., independent of) the human socio-political challenges that confront the remaining, extant ape populations. Promoting a realist-materialist approach, I advanced the agenda for theorising a *critical political ethology* as developed by York and Longo (2017) to: (a) take seriously the energetic and material exchanges between organism and environment; and (b) align dialectical principles with the emergence of an extended evolutionary synthesis (EES). In the analysis of detailed examples, specifically hylobatid community dynamics, behavioural differences in the genus *Pan*, and orangutan social flexibility, I demonstrated the potential of this approach. Beyond a simple call to reject reductionist thinking, I contend that the observed patterns of variability within and between hominoid taxa are simultaneously shaped by, and act as shaping factors of, evolutionary processes. This perspective aligns well with the tenets of the EES, namely: variation arising via constructive processes, and reciprocal causality within multiple channels of inheritance. To repeat a critical conclusion here: I see this emergence of behavioural variation as broadly similar to the expression of "reaction norms", where socioecological context dependency selects for behavioural ranges rather than singular types. Niche construction and the opening-up of multiple channels of inheritance provide a more dynamic, integrated set of evolutionary processes when compared to a more limited genetic response to socioecological determinants.

In all of the examples presented in Chapter 3, we see the role that human impacts can play in shaping contemporary ape socioecology. While apes have revealed their ability to adjust behaviourally to minor shifts in ecological and demographic conditions via social innovation and learning, they will continue to struggle in response to sweeping and swift disturbances to their habitats and/or population structures. Agricultural intensification and infrastructural development not only degrade and fragment habitats, but they also increase the risk to ape populations from illegal hunting, trade and the transmission of human diseases to wild apes (i.e., zooanthroponoses) (Arcus Foundation 2018). Finally, even the perspectives of researchers and their commitments to research paradigms can shape the ways in which data are collected and interpreted. In my own research, along with colleagues Megan Selby and Stefano Longo, we used a political ecology framework to understand how an endangered gibbon species is made "meaningful", but in different ways, by various stakeholders at local, regional and international scales (Malone et al. 2014). As we theorise ape sociality and contemplate research-informed conservation interventions, we do need to remain cognisant of congruencies and dissociations between the apes in the world and the apes in our heads.[6] With respect to conservation, development and narratives of human–wildlife conflict, political ecology serves as a response to, and critique of, the epistemological framing of biodiversity loss as primarily a locally induced crisis requiring interventionist policies designed to manage human–environment relationships (Escobar 1998; Carrier & West 2009; Malone et al. 2014: 41). Recognising *both* the material *and* symbolic importance of apes should form the basis for understanding how humans have impacted ape population histories, and therefore serve as a foundation for human–ape coexistence (York & Longo 2017).

Java and beyond

In Chapter 4, I took an in-depth look at the Indonesian island of Java from a variety of perspectives, including: biogeography, the political economy and ecology of natural resource management, and, importantly, the state of the island's primate communities. Drawing on my experience across various projects and sites, I have articulated the challenge (or indeed the impossibility) of separating the realms of the "natural" from that of the "cultural". Simultaneous attention to both socioecological data and ethnographic insights continue to reveal the complexity and nuances of primate life and human livelihoods in Java. Understanding these interwoven layers, and working productively towards sustainable outcomes for both human and nonhuman primates, is clearly the purview of an engaged and ethically grounded, anthropological primatology. As such, taking advantage of the many tools available in anthropology's broad toolkit is an utmost imperative. As discussed at length in the chapter, endangered silvery gibbons (the island's only extant ape) demonstrate a response sensitivity to the (nearly ubiquitous) presence of humans. Therefore, effective research and conservation activities depend upon our continued elucidation of both gibbon behavioural ecology and human activities within modified habitats. This applies equally to the remaining population strongholds, both outside of and within protected areas such as national parks. Certainly, the situation on Java is distinct, though it is not unique, and similar dynamics (and their underlying drivers) could be identified for many of the world's living primate species. I am confident that many of the insights from Java – derived from the dynamic interplay of cultural, ecological and historical forces – will be broadly applicable to primate populations on the brink, the world over.

Regardless of whether we identify ourselves as anthropologists, primatologists or tropical ecologists, we bemoan the scope and scale of humankind's alteration of our planet's ecosystems. At the same we are well placed to address what these alterations mean for species' ability to survive rapid environmental change (Behie et al. 2019). As the case studies from Java make clear, the social and natural processes are intertwined on many levels. For example, although extinction is an integral part of the evolutionary process, the recent exponential growth in the exploitation of global forest resources by humans demands our immediate attention, including the search for successful compromises, lest we intend to preside over a mass extinction event (Rose 2002; Estrada et al. 2017). Anthropogenic disturbance to ecological systems consists of a complex web of interactions related to the expansion of human populations and the associated patterns of settlement, agriculture and resource extraction. Over time, the repercussions of exploitative or extractive activities at the local level can expand, impacting both biotic and abiotic processes on a regional or even global scale. Logging results in increases of solar penetration, tempera-ture and wind exposure, transforming the structure of floral and faunal com-munities in forested ecosystems. Deforestation fundamentally compromises the ability of soil to retain moisture and balance the effects of rainfall seasonality. Subsequently, the increased run-off and flow of soil nutrients to the sea via streams enhances algae growth and alters the ecology of coastal ecosystems, including coral reefs (Baldwin 2003). The corresponding processes of deforest-ation (especially in the case of coastal mangrove forests) and the degradation of coral reefs were implicated in the devastating loss of human life following the major tectonic event and tsunami along the northwest coastline of Sumatra on Boxing Day 2004 (Pakpahan 2005; Widiyanto et al. 2020). In light of these complexities, the expansion of biological and biogeographic models to include both long-term and recent human patterns of interaction may prove benefi-cial in assessing the immediate susceptibility of some nonhuman primate taxa (Cowlishaw & Dunbar 2000).

Primates in/on the verge

The issues raised in Chapter 5 deal with the lives of displaced apes, especially in relation to the diversifying localities and methodologies of our investigations. I have demonstrated that *the sites of primatological relevance* are increasingly those where nonhuman primates exist within the margins of nature and society. In describing studies of captive apes in zoological gardens, temporarily displaced apes transitioning through rehabilitation and reintroduction programmes, or apes on the margins of protected/un-protected areas, I contend that such sites provide opportunities to re-examine fundamental human–ape dynamics. In blurring the boundaries between "wild" and "captive" apes, and embracing human management as a co-constructed, *naturalcultural* space, we can oper-ationalise the relationship between in-situ and ex-situ primate populations to understand the reciprocal impacts (Braverman 2014: 53). Extending pri-matology (typically within the framework of the now mature sub-field of ethnoprimatology) to include managed settings enhances the ability to untangle the complex webs of ideas about primates, and how those culturally mediated ideas shape the lives of both managed and wild primates. Correspondingly, the incorporation of social science research methods and theory is becoming common practice in primate research and conservation. Specifically, I have discussed the value of obtaining *ethnographic insight*. As evidenced by the various case studies covered in Chapter 5, adopting ethnographic methods has propelled primatology towards taking seriously the symbolic, social and ecological agency of the actors within the human–alloprimate interface. In studies of primates in captivity, in the wild and in the "in-between", I have made clear the necessity of not only attending to the myriad ethical considerations which inevitably arise, but indeed centralising an ethically engaged primatology as a core com-ponent of contemporary praxis (Malone & Palmer 2014; Riley & Bezanson 2018; Riley 2020). In sum, the integration of diverse epistemologies allows us to understand that relationships to nonhuman primates are multidimensional, and, as such, reveals synergies with a dialectical approach to primatology.

Constructing the conditions of coexistence

Throughout the book, I have been casting a critical eye on the prevailing tenets and practices of primatology. In fusing dialectical principles with progressive trends in the discipline, I have formed a basic framework for incorporating historical contingency and contradiction into the mode by which we theorise and research the human–ape interface. However, we stand at a critical juncture where our actions (or inactions) will determine the fate of our fellow hominoids. Our theorising and research inform the nature of our interventions, most commonly glossed as "primate conservation". Indeed, an explicit aspect of this book's rationale is to shape the way primatologists research and conserve nonhuman primates. Yet, we must not wait for further research findings or theoretical contemplations to commence the marshalling of resources required to preserve a future for the living apes.[7] The bold actions that are required will necessarily be underdetermined by predictive science. We know that "an optimal conservation strategy for primates is to protect as much heterogeneous habitat as possible" (Malone et al. 2010: 783). We know this without requiring further habituation of primate groups.[8] Habituation draws intensively upon time and resources, and also increases some risk factors (e.g., exposure to zooanthroponoses) for the animals under study. Research-focused field primatologists can still play a role in obtaining species richness and abundance data via survey work, and through the non-invasive monitoring of primate diets and demographics. We need a sea change with respect to public opinion and the types of collaborative partnerships that will allow us to reframe our responsibilities to endangered apes. And herein lies a role for the dialectical primatologist, and other activist-oriented academics. In the section that follows, I offer a synthesis by examining two polemical and dichotomous positions: "people versus protected areas", and "seeing zoos as either conservation partners or impediments". The perpetuation of these dichotomies, and the neglect in viewing them as a dialectic, in my opinion, maintains barriers to efficacious actions, and therefore need to be overcome in order to advance an agenda of hominoid coexistence.

Myth and reality, or the reality of myth?

Western perceptions regarding the preservation of a "pristine nature" produced a protectionist conservation ethic that sought to exclude human settlements and subsistence activities. In this vein, protected areas were established and managed, often by national governments, on customary lands. Frequently, such actions ignored long histories of human occupancy and use, leading to the displacement and disaffection of local peoples (Neumann 1998; Mombeshora & Le Bel 2009; Wade 2020). With tensions emerging from loss of land use rights, a counterbalancing trend emerged whereby attempts are made to link conservation with the health, educational and economic needs of local people (Caldecott 1996; Noss 1997; Hill 2002). In his 1999 book *Myth and Reality in the Rainforest*, John Oates, informed by decades of experience in Nigeria and Ghana, articulates an extremely critical view of such conservation strategies. Specifically, Oates takes aim at international conservation organisations, finding it "unimaginable" that they "would be supporting projects in West Africa whose effects promoted human economic development at the expense of vanishing wildlife" (Oates 1999: 17). Instead, Oates promotes a strategy whereby well-resourced conservation organisations, as well as experienced researchers and naturalists, (re-)prioritise the protection of threatened nature – sufficiently justified by its intrinsic value – over the improvement of human material well-being. In one example, Oates points to a shift in priorities away from a "traditionally managed" protected area (Cross River National Park) to an integrated framework that seeks to develop economic opportunities for those residing adjacent to the park in exchange for the decreased exploitation of the forest and its animal inhabitants. In the end, Oates documents the erosion of tangible conservation goals (e.g., the protection of Nigeria's gorillas) by political and financial expediencies related to the development of, and concessions to, rural communities.

Oates is certainly an authoritative voice, but his critique represents both a classic framing of biodiversity loss via primarily local patterns of resource exploitation, and also the failure of integrated approaches. Again, the alternative strategy (favoured by Oates) is to intervene with protectionist policies that seek to manage the human–environment relationship by severely restricting human access to wildlife habitats. In some targeted settings, particularly when frequent and unpredictable movements of people preclude the possibility of working with "local communities", such exclusionist interventions may play a role. However, what if this particular framing overshadows and oversimplifies causes of habitat degradation, and allows blame to fall on localised peoples without adequately understanding the complexity of human–environment relationships (Brown 1998; Berkes 2004; Adams & Hutton 2007)? I (and many others) have suggested that anthropology's toolkit can be deployed to examine the complexities of human–environment–animal relationships by considering alternative explanations for biodiversity loss and habitat degradation (Malone et al. 2014). Whereas Oates refers to the idea that Indigenous peoples can potentially live in sustainable harmony with nature as a dangerous myth, Codding et al. (2014: 659) argue that casting "indigenous peoples as either intentional conservationists or environmental devastators oversimplify human-environment interactions". In fact, in their research with the Aboriginal, desert-dwelling Martu people in Western Australia, Codding and colleagues demonstrate convincingly that mammalian diversity benefits from an intermediate level of human interaction. Specifically, Aboriginal hunting practices and fire-management regimes facilitate range expansion and population density for some species, including hill kangaroo (*Macropus robustus*). Though not designed explicitly to be "conservation measures", the dynamics under-lying such epiphenomenal benefits problematise existing narratives in conser-vation management that portray humans as inherent destructors. Furthermore, the genesis and perpetuation of these dynamics reside in a belief system that "places people within – not apart from – ecological interactions" (Codding et al. 2014: 666).[9] Given that providing a total barrier between ourselves and wild ape populations is unrealistic, we may find more efficacious conservation strategies by seeking to mitigate and moderate human impacts on ape habitats. To do so will require us to draw upon information from the entirety of our knowledges, including both the scientific and humanistic perspectives of all parties concerned.

Which brings us to this section's rejoinder – *the reality of myth*. Chapter 4 established in detail how material and ecological factors, which shape engagements among people, forests and animals, emerge from the belief system of rural Sundanese people. Beyond Java, expanding theoretical frameworks and incorporating diverse methodologies provide the possibility to understand peoples' resource use patterns in relation to broader political, economic and cultural forces.[10] Doing so can inform conservation policies that are poten-tially more equitable, responsible and effective. They can also help us to under-stand how non-Western knowledge systems can lead to tensions with more traditional conservation management plans (Leblan 2016; Amir 2019). These perspectives can also be applied to some of the same West African forests and species discussed by Oates. In 2014–2015, Alison Wade examined the com-plex entanglements among humans, Cross River gorillas (*Gorilla gorilla diehli*) and Nigeria-Cameroon chimpanzees (*Pan troglodytes ellioti*) in the unprotected Mone-Oku Forest, Cameroon (Wade et al. 2019; Wade 2020). Obtaining both ethnographic and ecological insights, Wade's research demonstrates the need to incorporate a combination of factors ranging from the history of colonial management, the varying perceptions of gorillas and chimpanzees and the pre-sent importance of cacao (*Theobroma cacao*) cultivation into the analysis of the human–ape interface. Wade found that of utmost salience for local villagers is a desire to expand their cultivation of cacao. Through alterations in ranging and the potential for crop-foraging, the nature of interactions among humans and apes are far from static. Interestingly, local perceptions of the apes range from fear to tolerance – a variability partially explained by power imbalances between people and conservation agendas. Wade's (2020: 239) analysis of anthropogenic alterations revealed that:

> These alterations continue to the present day, where political history has shaped limited livelihood alternatives, thereby increasing the reliance on the forest that has remained a constant in a more recent history of "devel-opmental" neglect and isolation. But this situation is not static, future alterations to the forest are also subject to regional and international pol-itical and economic influences such as the increased worldwide demand for cacao. Future research should consider furthering our understanding

of these apes' positions within the wider ecosystem in addition to intra-actions with humans before seeking to expand protected areas to conserve these apes.[11]

With the interpenetration of ecological, political and economic forces, Wade's characterisation of the entangled and shared histories of apes and humans questions the positioning of local people (as in dominant conservation narratives) as the key threat to survival of these endangered apes. Extrapolating from these specific examples to the greater array of human–ape interfaces, I reject the false choice of *either* community development *or* wildlife protection. Instead, I see the rural and agriculturally engaged peoples of the world as potentially maintaining the most tangible connections to, and familiarity with, the natural world. It is their wisdom that we can least afford to marginalise. Indeed, where we find embers of species coexistence, we must provide oxygen. In this perspective, scientific and conservation narratives are transformed to incorporate more humanistic tones: becoming narratives that acknowledge and empower the holders of Indigenous knowledge and values (Smith 2013).

Captivity as constraint or catalyst?

Machery (2013: 57) observed, "only a few people – mostly scientists studying apes and people in Africa and Asia living near ape populations – have the opportunity to have any direct, sustained interaction with [wild] apes". For most, fleeting observations in a zoological garden will be the extent of their close encounters with apes. Zoological gardens, the world over, are epicentres of animal representation and exhibition to millions of visitors annually. Therefore, the material and representational implications of keeping apes in captivity are crucial pieces of the coexistence puzzle. The justifications and the messages that are communicated to broader swathes of society will shape the collective actions (both positive and negative) that may largely determine the fates of endangered apes. Developing an appreciation of natural biotas and an awareness of the plight of endangered species, as achieved via the educational and recreational opportunities provided in the zoo setting, are common features of the mission statements for zoological gardens. The best zoos become conscientious conservation participants by providing training and research opportunities, and through fundraising in support of research and conservation activities.[12] However, inherent within the zoo setting, and pertaining to the idea of constructing multispecies coexistence, are concerns regarding the implicit reinforcement of a detrimental anthropocentrism. Indictments of zoological gardens and their associated societal impacts emanate from a diverse array of scholars and critics. Malamud (1998: 250), for example, suggests that animal representations within the public domain potentially set the tone "for manifold other human practices that exploit animals and nature based upon principles of non-reciprocity", a by-product of power-based relationships between viewers and subjects. This paradox is best demonstrated by the educational messages of zoos themselves: that captivity is either inhumane, less-than- ideal, and/or amoral, for some of the animals *exhibited in the zoo* (Waldau 2001; Malone et al. 2010: 781).

Arguably the most meaningful human–animal relationships fostered in the zoo environment are between the animals and the dedicated husbandry staff. Interestingly, even in these most meaningful relationships, keepers can become fraught with moral uncertainty and contradiction. For example, in Ally Palmer's interviews with keepers at the Auckland Zoo, she uncovered perspectives on orangutan welfare and autonomy that demonstrate a powerful aspect of captivity: a potential for *empathy* (Palmer 2012; Palmer et al. 2016; Palmer & Malone 2018). As summarised in Palmer and Malone (2018: 838–839):

> Keepers felt conflicted by their roles, as they spoke of orangutans as "people" and sometimes explicitly voiced support for great ape rights. Keeping orangutans as "prisoners" was therefore, as one keeper put it, a "gray area" and a "tricky issue", since "they are almost human in a cage". Two of the six keepers found it particularly difficult to justify their roles, with one describing feeling "guilty, all the time" and the other explaining it "breaks our hearts" to see orangutans in cages. Yet keepers were ultimately able to justify their roles to themselves by either pragmatic reasoning ("They're here. We can't do anything about that realistically, so who better to look after them than me, or us, people who really, really care?"), or by invoking educational or ark justifications for zoos.

Empathy for apes is an essential element of relationships across species boundaries and could serve as a foundation for a widespread ethic of coexistence. However, the vital question that remains unanswered for me about the role of zoos is: *can they facilitate sufficient empathy among the zoo-going public so as to offset their inherent anthropocentrism?* On one hand, the very existence of zoos symbolises a mastery and domination of animals that lies in inherent contradiction with their mission to foster harmonious relationships with nature. On the other hand, while they exist, they could exert significant influence over the public framing of captivity for future generations. Instead of buffering visitors from the realities of ape extinctions by offering a view of a generic, exemplar "ape" (Machery 2013), zoos could be brutally honest about what they represent. Rather than promoting captive animals as ambassadors for their species, we should view the life of each and every zoo-housed ape as a tragic reminder of our collective failures and culpabilities. As the late John Berger wrote: "Everywhere animals disappear. In zoos they constitute the living monument to their own disappearance … the zoo is a demonstration of the relations between man [*sic*] and animal; nothing else" (1980: 26). Increasing our levels of empathy for animals, and seeing them as individuals, could activate the agency of animals and radically challenge a critical dichotomy – between us and them (Haraway 2006; Palmer et al. 2016; Kopnina 2017). Doing so could result in a major shift in discourses around humans' destructive impact on the natural world, since it would force us to expand "our concerns about the disappearance of an abstract category to include the concrete reality of death by starvation or disease or poaching of multitudes of feeling, thinking, relational individuals" (Smuts 2006: 125–126).

In conclusion

Throughout the book, I have tried to ask and answer some vexing questions about our research and conservation engagements with fellow members of the superfamily Hominoidea. Though certainly not the first to grapple with these topics, I have aimed to address them through a new lens. In recognising how the inseparability of the natural and the social impacts our science and our subjectivities, I see the potential for a dialectical approach to highlight the heterogeneity and complexity of ecological relationships. Fundamental to my perspective is an embrace of new approaches and conceptual developments that elevate integrative niche dynamics to a position of analytical primacy (e.g., Odling-Smee et al. 1996; Laland et al. 2001; Fuentes 2015). The environments of (ongoing) hominoid evolution are not merely external and passively inherited; apes actively select and transform their social and ecological niches. Moreover, as fellow hominoids (or "*ex*-apes", in Jonathan Marks' words), our ape research and conservation activities are contested and defended along both biological and cultural bases. While reductionist and non-dialectical approaches in primatology provide utility in terms of hypothesis generation and model formation, they can often veil consequential biological and social interpenetrations. For example, within a reductionist paradigm, the genetic proximity of chimpanzees and humans preconditions us towards a search for shared behavioural adaptations, regardless of the absence of any specific gene-trait linkages. In the realm of conservation, the biannual determination of the "World's 25 Most Endangered Primates" is an explicitly cultural and political process.[13] To repeat: both the study of apes, and primate conservation, are cultural endeavours.

Furthermore, by embracing the conceptual "anthropocene", and perhaps more importantly acknowledging the metrics of the "great acceleration" (i.e., the exponential growth in human population and consumption-levels, post-1950), we accept that anthropogenic alterations to the planet's land, oceanic and atmospheric systems interact in complex ways to impact biodiversity (Steffen et al. 2007, 2015). Of course, we too are subject to the effects of our actions. Engels (1966: 180) reminds us:

> Let us not, however, flatter ourselves overmuch on account of our human conquest over nature. For each such conquest takes its revenge on us. Each victory, it is true, has in the first place the consequences on which we counted, but in the second and third places it has quite different, unforeseen effects which only too often cancel out the first ... Thus at every step we are reminded that we by no means rule over nature like a conqueror

over a foreign people, like someone standing outside nature – but that we, with flesh, blood, and brain, belong to nature, and exist in its midst …

Of particular relevance to readers here, a heightened awareness of humankind's influence on primate communities has arguably altered the primatological endeavour *fundamentally and permanently*. In embracing the anthropocene, we recognise a present where "nature" is shaped and defined by human activity (Lorimer 2015; Malone et al. 2019). Will the future involve an untangling of our entanglements, a selective enhancement or rejection of our management over formerly wild places and species, or indeed an embrace of our multispecies, post-nature relationships and correspondingly, a shift in what we mean by "conservation"?

The challenges facing the apes are both immense and multifaceted. And yet, they are in fact a subset of the planet's overall problems with environmental degradation. Indeed, this environmental degradation continues at pace and on a massive scale. This should not dissuade us from taking action. In fact, we can feel empowered to know that our species' social history and future is linked with (though not entirely reducible to) our natural history, writ large. We can act to shape the direction of our biocultural evolutionary trajectory. And here is where the dialectical approach proves essential. Eloquently, Clark and York (2005: 21) state:

> Evolution does not entail a drive towards perfection, nor is there a perfect state of balance in existence (much less waiting to be discovered). Keeping things as they are is not an option, because change is a constant. But we must try to gain as much knowledge as we can regarding the conditions and processes of the world, so we can try to affect the course of change, to whatever degree is possible given historical constraints and conditions, in order to make a decent world possible for all life. The extent to which we succeed as social agents will determine the longevity of social history in the making, our relation to the natural world, and the potential for revolutionary change.

Following the tradition of Lewontin and Levins (2007: 10) in the need to "relate theory to real-world problems as well as the importance of theoretical critique", I recognise the tension between the philosophical nature of this book and my attempt to connect it to the urgent demands of the primate extinction crisis. While we all might agree that our fellow hominoids should benefit from protected, expansive habitats, we may disagree as to who should bear the costs. Marginalised local communities are not the producers of endangered apes, but rather both parties are products of inequitable and insatiable, wide-scale economic dynamics. With the appropriate balance of analysis and action, we can alter our politics, economies and behaviour to release the victims (human and nonhuman alike) from the worst effects of the present, oppressive system.

Notes

1 Behie et al. (2019: 5–6) argue that "acknowledging humankind's near ubiquitous impact on the environment may serve as a prerequisite for our active management of diverse, species-rich habitats (rather than attempts to protect unspoilt, 'natural' wilderness)". Others, however, suggest that the terminology of the anthropocene presents both practical and ideological complications, as well as political implications (Caro et al. 2012; Sayre 2012). Caro et al. (2012) point to four potential negative consequences of thinking that *humans have altered everything*, including: 1) increasing our tolerance for highly manipulative rewilding campaigns; 2) deprioritising initiatives to protect nearly intact ecosystems; 3) an enhanced ability for governments to further land-use projects if habitats are already viewed as degraded; and 4) risking the spread of public pessimism and loss of monetary support for conservation agendas if the whole of nature is perceived to be altered by humans.

2 Overall, 75% of the world's primates are in steady decline and 60% are threatened with extinction (Estrada et al. 2017). The status of primates reflects a more general trend among terrestrial vertebrates, where at least 515 species are considered to be "on the brink" (i.e., species with fewer than 1,000 individuals remaining) (Ceballos et al. 2020).

3 "The *Monthly Review* … was and is Marxist, but did not hew to the party line or get into sectarian struggles". – Current editor John Bellamy Foster (2000–present) to *The New York Times* on 2 March 2004 following the death of one the magazine's founding editors, Paul Sweezy.

4 Richard Levins (1996: 103–104) draws attention to what he refers to as "the dual nature of science":

> On the one hand, it really does enlighten us about our interactions with the rest of the world, producing understanding and guiding our actions. We really have learned a great deal about the circulation of the blood, the geography of species, the folding of proteins, and the folding of the continents. We can read the fossil records of a billion years ago, reconstruct the animals and climates of the past and the chemical compositions of the galaxies, trace the molecular pathways of neurotransmitters and the odour trails of ants. And we can invent tools that will be useful long after the theories that spawned them have become quaint footnotes in the history of knowledge. On the other hand, as a product of human activity, science reflects the conditions of its production and the viewpoints of its producers or owners. The agenda of science, the recruitment and training of some and the exclusion of others from being scientists, the strategies of research, the physical instruments of investigation, the intellectual framework in which problems are formulated and results interpreted, the criteria for a successful solution to a problem, and the conditions of application of scientific results are all very much a product of the history of the sciences and associated technologies and of the societies that form and own them. The pattern of knowledge and ignorance in science is not dictated by nature but is structured by interest and belief. We easily impose our own social experience onto the social lives of baboons, our understanding of orderliness in business, implying a hierarchy of controllers and controlled, onto the regulation of ecosystems and nervous systems. Theories, supported by mega-libraries of data, often are systematically and dogmatically obfuscating.

5 Marks (2012: 101) rightfully points out that:

> we were not always apes; we became apes as a consequence of the cultural privilege accorded to genetic data and approaches at the end of the 20th century. We became apes as a dialectical relationship of descent and modification became replaced by a reductive view, in which descent (which genetics reveals well) supersedes modification (which genetics does not reveal well).

6 Rothfels (2002) demonstrates how descriptions of apes (gorillas in this example) align with particular historical moments. From Paul Du Chaillu's mid-19th-century explorations and hunting encounters in equatorial Africa where the gorilla is portrayed as a menacing threat, to Alexander Sokolowsky's early-20th-century psychological analysis of the introspective and intellectual gorilla, every human generation places its animals within its particular historical moment. According to Rothfels (2002: 4–5):

> the examples of Sokolowsky and Du Chaillu demonstrate that, while individual gorillas have unique identities, our access to those identities is constrained by the mediated nature of their presence in our historical record. As far as the historical record is concerned, gorillas of the past do not represent themselves; rather Du Chaillu's "gorilla", Sokolowsky's "gorilla", and even Dian Fossey's "gorilla" are entities inextricably bound by particular human contexts and human interpretations. That is, there is an inescapable difference between what an animal *is* and what people *think* an animal is [emphasis original]. In the end, an animal or species is as much a constellation of ideas (for example, vicious, noble, intelligent, cruel, caring, brave) as anything else.

7 More research, science and technology, does not necessarily produce the attributes we require to overcome the present extinction crisis, namely: motivation, optimism, compassion and proactive social change (Bekoff & Bexell 2010; Bekoff 2013).

8 To clarify, and as stated in Malone et al. (2010: 783):

> As we are not renouncing all primate field studies (nor announcing our retirements), let us be clear with what we are suggesting. We are suggesting that an ethical primatology is found neither in the thesis of a research-dominated engagement with nonhuman primates, nor the antithesis of a purely human-istic engagement, but in a higher synthesis. It is found in a higher synthesis that combines the verities of both. Continuing to study existing populations of habituated primates may be wise on both conservation and research related grounds. In fact, discontinuing such research may in fact be uneth-ical in some cases – exposing habituated primates to an under-employed and under-supported human population now accustomed to the resources that primatologists provide.

9 See Bliege Bird et al. (2013) for additional documentation of the material and ecological effects of the *Jukurrpa*, or ecological and cosmological knowledge as obtained from the Dreamtime ancestors.

10 As an example, the traditional forest management practices and norms (including taboos) of Iban (Dayak) communities have been shown to buffer Bornean orang-utan (*Pongo pygmaeus*) from the full effects of unrestrained hunting (Wadley et al. 1997; Wadley & Colfer 2004; Yuliani et al. 2018).

11   Alison Wade, following Jost Robinson and Remis (2018), refers the relational nature of human–alloprimate *intra-* (rather than *inter-*) actions. For multispecies assemblages, Jost Robinson and Remis (2018: 782), citing Karen Barad (2003, 2007), Donna Haraway (2013) and others, find an enhanced understanding by focusing on intra-actions, or "a diffractive process in which humans or alloprimates, or any combination of living and/or nonliving entities are changed, and continue to be changed, as constituent parts of a holistic system".

12   Auckland Zoo is a long-term supporter of conservation initiatives both in New Zealand and abroad, including projects in primate-range countries (e.g., Indonesia). For example, in 2017/2018 a grant of $35,000 NZD was awarded to support the Sumatran Orangutan Conservation Project (SOCP). In the nearly 20-year partnership, the Auckland Zoo has awarded over a half a million dollars to the SOCP to support rehabilitation, law enforcement and education efforts to protect the Sumatran orangutan (*Pongo abelii*). See Auckland Zoo's Field Conservation Annual Report 2017–2018 at https://issuu.com/aucklandzoo/docs/auckland_zoo_field_conservation_ann. While this represents a sustained and sizeable commitment, the financial investment is modest relative to the investment in Auckland Zoo's new $60 million dollar "South East Asia Jungle Track". This immersive development features an enriched habitat for orangutan, siamang, tigers and other SE Asian fauna, as well as a "stunning new café/function venue overlooking our lake" for zoo patrons. See: www.aucklandzoo.co.nz/south-east-asia.

13   The list, co-produced by the Primate Specialist Group (IUCN/SSC) and Global Wildlife Conservation on a biennial basis attracts significant media attention. While formulated on an empirical basis and in open consultation with the primatological community, the list is not intended to reflect the 25 most endangered primate species in an absolute sense, but rather explicitly (and strategically) highlights species in order to raise awareness and mobilise urgent conservation action. The 10th iteration (2018–2020) was constructed at the 27th Congress of the International Primatological Society in Nairobi, Kenya, in August 2018.

References

Adams WM, Hutton J. 2007. People, parks and poverty: Political ecology and biodiversity conservation. *Conservation and Society* 5(2): 147–183.

Amir APH. 2019. Who knows what about gorillas? Indigenous knowledge, global justice, and human–gorilla relations. *IK: Other Ways of Knowing* 5: 1–40.

Arcus Foundation. 2018. *Infrastructure Development and Ape Conservation* (Vol. 3). Cambridge: Cambridge University Press.

Baldwin JR. 2003. *Reconceiving Wealth for Geographic Analysis*. PhD Dissertation, University of Oregon, Eugene, OR.

Barad K. 2003. Posthumanist performativity: Toward an understanding of how matter comes to matter. *Signs: Journal of Women in Culture and Society* 28(3): 801–831.

Barad K. 2007. *Meeting the Universe Halfway: Quantum Physics and the Entanglement of Matter and Meaning*. Durham, NC: Duke University Press.

Behie A, Teichroeb J, Malone N. (Eds.). 2019. Changing priorities for Primate Conservation and Research in the Anthropocene. In Behie A, Teichroeb J, Malone N. (Eds.), *Primate Research and Conservation in the Anthropocene* (Vol. 82, pp. 1–13). Cambridge: Cambridge University Press.

Bekoff M. 2013. Who lives, who dies, and why? In Corbey R, Lanjouw A. (Eds.), *The Politics of Species: Reshaping our Relationships with Other Animals* (pp. 15–26). Cambridge: Cambridge University Press.

Bekoff M, Bexell SM. 2010. Ignoring nature: Why we do it, the dire consequences, and the need for a paradigm shift to save animals, habitats, and ourselves. *Human Ecology Review* 17(1): 70–74.

Berger J. 1980. *About Looking*. New York: Vintage.

Berkes F. 2004. Rethinking community-based conservation. *Conservation Biology* 18(3): 621–630.

Bliege Bird R, Tayor N, Codding BF, Bird DW. 2013. Niche construction and Dreaming logic: Aboriginal patch mosaic burning and varanid lizards (*Varanus gouldii*) in Australia. *Proceedings of the Royal Society B: Biological Sciences* 280(1772): 20132297.

Braverman I. 2014. Conservation without nature: The trouble with in situ versus ex situ conservation. *Geoforum* 51: 47–57.

Bronikowski AM, Cords M, Alberts SC, Altmann J, Brockman DK, Fedigan LM, Pusey A, Stoinski T, Strier KB, Morris WF. 2016. Female and male life tables for seven wild primate species. *Scientific Data* 3(1): 1–8.

Brown K. 1998. The political ecology of biodiversity, conservation and development in Nepal's Terai: Confused meanings, means and ends. *Ecological Economics* 24(1): 73–87.

Caldecott JO. 1996. *Designing Conservation Projects*. Cambridge: Cambridge University Press.

Caro T, Darwin J, Forrester T, Ledoux-Bloom C, Wells C. 2012. Conservation in the Anthropocene. *Conservation Biology* 26(1): 185–188.

Carrier JG, West P. 2009. Introduction. In Carrier JG, West P. (Eds.), *Virtualism, Governance, and Practice: Vision and Execution in Environmental Conservation* (pp. 1–23). New York, NY: Berghahn Books.

Ceballos G, Ehrlich PR, Raven PH. 2020. Vertebrates on the brink as indicators of biological annihilation and the sixth mass extinction. *Proceedings of the National Academy of Sciences* 117(4): 13596–13602.

Clark B, York R. 2005. Dialectical nature. *Monthly Review* 57(1): 13–22.

Codding BF, Bliege Bird R, Kauhanen PG, Bird DW. 2014. Conservation or co-evolution? Intermediate levels of aboriginal burning and hunting have positive effects on kangaroo populations in Western Australia. *Human Ecology* 42(5): 659–669.

Cowlishaw G, Dunbar R. 2000. *Primate Conservation Biology*. Chicago, IL: University of Chicago Press.

Engels F. 1966. *Dialectics of Nature*. Moscow: Progress Publishers.

Escobar A. 1998. Whose knowledge, whose nature? Biodiversity, conservation, and the political ecology of social movements. *Journal of Political Ecology* 5: 53–82.

Estrada A, Garber PA, Rylands AB, Roos C, Fernandez-Duque E, Di Fiore A, … & Rovero F. 2017. Impending extinction crisis of the world's primates: Why primates matter. *Science Advances* 3(1): e1600946.

Foster JB. 2000. *Marx's Ecology: Materialism and Nature*. New York, NY: Monthly Review Press.

Fuentes A. 2015. Integrative anthropology and the human niche: Toward a contemporary approach to human evolution. *American Anthropologist* 117(2): 302–315.

Fuentes A, Baynes-Rock M. 2017. Anthropogenic landscapes, human action and the process of co-construction with other species: Making anthromes in the Anthropocene. *Land* 6(1): 15.

Haraway D. 2006. Encounters with companion species: Entangling dogs, baboons, philosophers, and biologists. *Configurations* 14(1): 97–114.

Haraway DJ. 2013. *When Species Meet* (Vol. 3). Minneapolis, MN: University of Minnesota Press.

Hill CM. 2002. Primate conservation and local communities – ethical issues and debates. *American Anthropologist* 104(4): 1184–1194.

Hockings KJ, McLennan MR, Carvalho S, Ancrenaz M, Bobe R, Byrne RW, ... & Wilson ML. 2015. Apes in the Anthropocene: Flexibility and survival. *Trends in Ecology & Evolution* 30(4): 215–222.

Jablonski NG. 2005. Primate homeland: Forests and the evolution of primates during the Tertiary and Quaternary in Asia. *Anthropological Science* 113: 117–122.

Jost Robinson CA, Remis MJ. 2018. Engaging holism: Exploring multispecies approaches in ethnoprimatology. *International Journal of Primatology* 39(5): 776–796.

Kelley J. 1997. Paleobiological and phylogenetic significance of life history in Miocene hominoids. In Begun DR, Ward CV, Rose MD. (Eds.), *Function, Phylogeny, and Fossils: Miocene Hominoid Origins and Adaptations* (pp. 173–208). New York, NY: Plenum Press.

Kopnina H. 2017. Beyond multispecies ethnography: Engaging with violence and animal rights in anthropology. *Critique of Anthropology* 37(3): 333–357.

Kramer KL. 2019. How there got to be so many of us: The evolutionary story of population growth and a life history of cooperation. *Journal of Anthropological Research* 75(4): 472–497.

Laland KN, Odling-Smee FJ, Feldman MW. 2001. Niche construction, ecological inheritance, and cycles of contingency in evolution. In Oyama S, Griffiths PE, Gray RD. (Eds.), *Cycles of Contingency: Developmental Systems and Evolution* (pp. 117–126). Cambridge, MA: MIT University Press.

Laurance WF, Sayer J, Cassman KG. 2014. Agricultural expansion and its impacts on tropical nature. *Trends in Ecology and Evolution* 29: 107–116.

Leblan V. 2016. Territorial and land-use rights perspectives on human–chimpanzee-elephant coexistence in West Africa (Guinea, Guinea-Bissau, Senegal, nineteenth to twenty-first centuries). *Primates* 57(3): 359–366.

Levins R. 1996. Ten propositions on science and antiscience. *Social Text* 46/47: 101–111.

Levins R, Lewontin R. 1985. *The Dialectical Biologist*. Cambridge, MA: Harvard University Press.

Lewontin R, Levins R. 2007. *Biology Under the Influence: Dialectical Essays on Ecology, Agriculture and Health*. New York, NY: Monthly Review Press.

Littleton J. 2005. Fifty years of chimpanzee demography at Taronga Park Zoo. *American Journal of Primatology* 67(3): 281–298.

Longo SB, Malone N. 2006. Meat, medicine, and materialism: A dialectical analysis of human relationships to nonhuman animals and nature. *Human Ecology Review* 13: 111–121.

Lorimer J. 2015. *Wildlife in the Anthropocene: Conservation after Nature*. Minneapolis, MN: University of Minnesota Press.

Machery E. 2013. Apeism and racism. In Corbey R, Lanjouw A. (Eds.), *The Politics of Species: Reshaping Our Relationships with Other Animals* (pp. 53–66). Cambridge: Cambridge University Press.

Malamud R. 1998. *Reading Zoos: Representations of Animals and Captivity*. New York, NY: New York University Press.

Malone NM, Fuentes A, White FJ. 2010. Ethics commentary: Subjects of knowledge and control in field primatology. *American Journal of Primatology* 72(9): 779–784.

Malone N, Palmer A. 2014. Ethical issues within human–alloprimate interactive zones. In *Engaging with Animals: A Shared Existence* (pp. 21–38). Sydney: University of Sydney Press.

Malone N, Selby M, Longo SB. 2014. Political and ecological dimensions of silvery gibbon conservation efforts: An endangered ape in (and on) the verge. *International Journal of Sociology* 44(1): 34–53.

Malone N, Teichroeb JA, Behie AM. 2019. Research(ers) and conservation(its) in the Anthropocene. In Behie A, Teichroeb J, Malone N. (Eds.), *Primate Research and Conservation in the Anthropocene* (Vol. 82, pp. 281–284). Cambridge: Cambridge University Press.

Marks J. 2012. Why be against Darwin? Creationism, racism, and the roots of anthropology. *American Journal of Physical Anthropology* 149(S55): 95–104.

Mombeshora S, Le Bel S. 2009. Parks-people conflicts: The case of Gonarezhou National Park and the Chitsa community in south-east Zimbabwe. *Biodiversity and Conservation* 18(10): 2601–2623.

Neumann RP. 1998. *Imposing Wilderness: Struggles over Livelihood and Nature Preservation in Africa*. Berkeley: University of California Press.

Noss AJ. 1997. Challenges to nature conservation with community development in central African forests. *Oryx* 31(3): 180–188.

Oates JF. 1999. *Myth and Reality in the Rain Forest: How Conservation Strategies are Failing in West Africa*. Berkeley: University of California Press.

Odling-Smee FJ, Laland KN, Feldman MW. 1996. Niche construction. *The American Naturalist* 147(4): 641–648.

Pakpahan H. 2005. Mangrove, pemecah kekuatan tsunami. *Habitat* 5(1): 12–15.

Palmer A. 2012. *Keeper/Orangutan Interactions at Auckland Zoo: Communication, Friendship, and Ethics Between Species*. MA Thesis, University of Auckland.

Palmer A, Malone N. 2018. Extending ethnoprimatology: Human–alloprimate relationships in managed settings. *International Journal of Primatology* 39(5): 831–851.

Palmer A, Malone N, Park J. 2016. Caregiver/orangutan relationships at Auckland Zoo: Empathy, friendship, and ethics between species. *Society & Animals* 24(3): 230–249.

Riley EP. 2020. *The Promise of Contemporary Primatology*. New York, NY: Routledge.

Riley EP, Bezanson M. 2018. Ethics of primate fieldwork: Toward an ethically engaged primatology. *Annual Review of Anthropology* 47: 493–512.

Ritchie C, Knight A. 2016. Asymmetric flows, critical zones, and zero-carbon citizens: A report on: "How to think the anthropocene?", November 2015, Paris, France. *Critique of Anthropology* 36(2): 212–220.

Rose AL. 2002. Conservation must pursue human–nature biosynergy in the era of social chaos and bushmeat commerce. In Fuentes A, Wolfe L. (Eds.), *Primates Face to Face: Conservation Implications of Human-Nonhuman Primate Interconnections* (pp. 208–239). Cambridge: Cambridge University Press.

Rothfels N. 2002. *Savages and Beasts: The Birth of the Modern Zoo*. Baltimore, MD: The Johns Hopkins University Press.

Sayre NF. 2012. The politics of the anthropogenic. *Annual Review of Anthropology* 41: 57–70.

Smith LT. 2013. *Decolonizing Methodologies: Research and Indigenous Peoples*, 2nd ed. London: Zed Books Ltd.

Smuts B. 2006. Between species: Science and subjectivity. *Configurations* 14(1): 115–126.

Steffen W, Broadgate W, Deutsch L, Gaffney O, Ludwig C. 2015. The trajectory of the Anthropocene: The great acceleration. *The Anthropocene Review* 2(1): 81–98.

Steffen W, Crutzen PJ, McNeill JR. 2007. The Anthropocene: Are humans now overwhelming the great forces of nature. *AMBIO: A Journal of the Human Environment* 36(8): 614–621.

Temerin LA, Cant JG. 1983. The evolutionary divergence of Old World monkeys and apes. *The American Naturalist* 122(3): 335–351.

van Schaik C. 2013. The costs and benefits of flexibility as an expression of behavioural plasticity: A primate perspective. *Philosophical Transactions of the Royal Society B: Biological Sciences* 368(1618): 20120339.

Wade AH. 2020. *Shared Landscapes: The Human-Ape Interface within the Mone-Oku Forest, Cameroon.* PhD Thesis, University of Auckland, New Zealand.

Wade A, Malone N, Littleton J, Floyd B. 2019. Uneasy neighbours: Local perceptions of the Cross River gorilla and Nigeria-Cameroon chimpanzee in Cameroon. In Behie A, Teichroeb J, Malone N. (Eds.), *Primate Research and Conservation in the Anthropocene* (Vol. 82, pp. 52–73). Cambridge: Cambridge University Press.

Wadley RL, Colfer CJP. 2004. Sacred forest, hunting, and conservation in West Kalimantan, Indonesia. *Human Ecology* 32(3): 313–338.

Wadley RL, Colfer CJP, Hood IG. 1997. Hunting primates and managing forests: The case of Iban forest farmers in Indonesian Borneo. *Human Ecology* 25(2): 243–271.

Waldau P. 2001. Inclusivist ethics. In Beck BB, Stoinski TS, Hutchins M, Maple TL, Rowan A, Stevens EF, Arluke A. (Eds.), *Great Apes and Humans: The Ethics of Coexistence* (pp. 295–312). Washington, DC: Smithsonian Institution Press.

Wich SA, Utami-Atmoko SS, Setia TM, Rijksen HD, Schürmann C, Van Hooff JARAM, van Schaik CP. 2004. Life history of wild Sumatran orangutans (*Pongo abelii*). *Journal of Human Evolution* 47(6): 385–398.

Widiyanto W, Hsiao SC, Chen WB, Santoso PB, Imananta RT, Lian WC. 2020. Run-up, inundation, and sediment characteristics of the 22 December 2018 Sunda Strait tsunami, Indonesia. *Natural Hazards and Earth System Sciences* 20(4): 933–946.

York R, Longo SB. 2017. Animals in the world: A materialist approach to sociological animal studies. *Journal of Sociology* 53(1): 32–46.

Yuliani EL, Adnan H, Achdiawan R, Bakara D, Heri V, Sammy J, Agus Salim M, Sunderland T. 2018. The roles of traditional knowledge systems in orang-utan *Pongo* spp. and forest conservation: A case study of Danau Sentarum, West Kalimantan, Indonesia. *Oryx* 52(1): 156–165.

Epilogue

The view from Aotearoa New Zealand

Manaaki whenua, manaaki tangata, haere whakamua
[Care for the land, care for people, move forward]

(Māori whakataukī (proverb))

We need another and a wiser and perhaps a more mystical concept of animals. Remote from universal nature and living by complicated artifice, man [sic] in civilisation surveys the creature through the glass of his knowledge and sees thereby a feather magnified and the whole image in distortion. We patronise them for their incompleteness, for their tragic fate for having taken form so far below ourselves. And therein do we err. For the animal shall not be measured by man. In a world older and more complete than ours, they move finished and complete, gifted with the extension of the senses we have lost or never attained, living by voices we shall never hear. They are not brethren, they are not underlings: they are other nations, caught with ourselves in the net of life and time, fellow prisoners of the splendour and travail of the earth.

Naturalist Henry Beston, from *The Outermost House* (1928)

I hadn't planned on writing the majority of this book sheltered within the relative isolation of my household amidst a global pandemic. In Aotearoa New Zealand, my adopted homeland, the well-conceived moves to shut down the society came hard and fast. Like many academics and primate fieldworkers, the sudden cancellation of travel plans altered my research agenda to an appreciable degree. Furthermore, with the release of new IUCN guidelines for research protocols during and after the pandemic, including a call to suspend great ape tourism and reduce field research, it occurred to me that the very future of primatology (at least in the short term) could be transforming in real time.[1] How might an increased emphasis on the health risks of zoonotic diseases and the global trafficking of wild animals shift our roles and responsibilities as primatologists? How might a reduced presence at our field sites impact the populations and communities with which we've frequently engaged? Would non-invasive or indirect methods of data collection replace more traditional observational methods, thereby decreasing inter-species proximity and the

DOI: 10.4324/9780367211349-7

need to habituate study animals to human presence? Primatologists are already demonstrating leadership in addressing such questions, and offering tangible recommendations for enhancing the health and safety of both human and non-human primate communities (Lappan et al. 2020). Moving beyond the practical, have the sudden temporal and spatial shifts to many human—nonhuman primate interfaces (e.g., sans tourists and international researchers) created space to conceptualise a different relationship between humans and nonhuman animals; between observer and the observed?

In this regard, it is important to acknowledge variation in conceptual baselines. As a discipline, with clearly established (yet varying) European, American and Japanese traditions, primatology is heavily influenced by the Northern Hemisphere (Jost Robinson & Remis 2018; Riley 2020). The Southern Hemisphere in general, and Australia/New Zealand specifically, has been under-represented in primatology despite reasonably well-funded tertiary education sectors and a relative proximity to wild primate range states in Asia. Primatologists, with identities consisting of ascribed (e.g., ethnicity, gender and socioeconomic status) and acquired (e.g., education, training and experiential) characteristics, as well as biases with respect to their conception of research, nature and conservation, play a role in shaping both the science of primate studies and emerging conceptions of interspecies relationships (Haraway 1989; Latour 2000; Fuentes 2011; Malone et al. 2014). With long-range international travel largely paused, it seems as good a time as any to reflect upon the primatological history of one's "local" – in my case, Australasia. Arguably, the most important contributions to primatology in the region were made by Professor Colin Groves (Behie & Oxenham 2015).[2] Moving to Canberra in 1973/1974 from England to begin a long and illustrious career at the Australian National University (ANU), Colin's passing in late-2017 represents a major loss. Colin trained numerous students, many of whom (e.g., Erik Meijaard, Ben Rawson and Anton Nurcahyo) have gone on to make major contributions in primate research and conservation. Equally noteworthy is the legacy of Charles Oxnard and his leadership of Anatomy and Human Biology at the University of Western Australia (UWA). Indeed, Charles is the founding president of the Australasian Society for Human Biology. Today primatology in Australia is flourishing under the leadership of Alison Behie and her students at ANU, as well as Deb Judge and Cyril Grueter at UWA. Here in New Zealand, primatology is still maturing. To date, New Zealand's contributions to primatology come from: Christina Campbell, originally from Christchurch, who received her PhD in Anthropology from the University of California (Berkeley) and specialises in spider monkey behaviour and ecology; Hazel Chapman at the University of Canterbury who has trained students in evolutionary ecology (including primates) in West Africa; and Weihong Ji at Massey University who studies vertebrate behavioural ecology, including snub-nosed monkeys in China. At the University of Auckland, biological anthropologist John Allen supervised a small number of primatological projects in the 1990s, including Sharon Watt's doctoral thesis on the socioecology of red colobus monkeys. Later, Brazilian primatologist Jean Boubli held a lectureship in anthropology from 2005 to 2009 before leaving to become a programme director for the Wildlife Conservation Society. I arrived in 2010 and began building a programme with an emphasis on ethnoprimatology. In 2020, Alison Wade received a doctoral degree for her ethnoprimatological research on the human–ape interface in Cameroon, a first of its kind for New Zealand.

Since immigrating to New Zealand over a decade ago, I have witnessed government and societal responses to several major crises: devastating earthquakes in Christchurch and Kaikoura; a fatal tragedy in a South Island mine; the country's deadliest mass shooting targeting worshippers inside of mosques (again in Christchurch); and the eruption of Whakaari – a volcanic tourist site in the Bay of Plenty. And now the COVID-19 pandemic. Generally speaking, in all of these events there has been a certain pragmatic and above all *collective* approach to the responses and actions. From the highest levels of government down to the interpersonal dynamics between fellow New Zealanders, there is a level of compassion amidst the politics that stands in stark contrast to what I had grown accustomed to during my previous decades of time spent in the United States. I do not mean to portray Aotearoa New Zealand as a utopian society free from the ills of structural inequality, as, indeed, contemporary life extends upon a colonial history and fraught relations with Indigenous people.[3] That said, the nurturance of bicultural societal relations is evident in the relatively widespread and supported use of Māori concepts such as: *manaakitanga* (the process of showing respect, generosity and care for others); *whanaungatanga* (the sense of familial or reciprocal relationships); and *kaitiakitanga* (a guardianship role for the environment, knowledge and traditions). Though imperfect, what impresses me are the embers of a shared purpose – an ideological cohesiveness – that can be stoked during times of crisis. Elements of this cohesiveness can be found in a sense of intergenerational awareness of place and relationships, built upon a tangible connection with the land. From these foundations, empathetic leadership can be supported. Whether it is honouring the sacrifices of the ANZAC soldiers at Gallipoli, or the swift and nearly unanimous vote to ban military-style assault weapons within weeks of the Christchurch massacre,[4] common understandings of justice and injustice – of a shared humanity – can be maintained. The unification of people towards a common goal is an effective antidote for the divisive forces of our times. In my opinion such effective, collective responses (which I argue are essential for constructing the conditions for hominoid coexistence) demonstrate the real possibility of re-thinking and re-making relationships.

I am also influenced, as I conclude this book, by the global reaction (rightfully appalled, but tragically unsurprised) to the killing of George Floyd, Breonna Taylor and other Black Americans, as well as the horrendously inequitable impacts of COVID-19 on racialised groups in the United States.[5] In solidarity, I recognise the lived experiences of minority communities subjected to oppression. As a biological anthropologist, I know that race is not rooted in biology, but in the policies and practices (racism) of those who seek to dominate others. Simultaneously, I must acknowledge the historical and contemporary place of anthropology in contributing to both the scientific and public discourse on race and racism – both the good and the bad.[6] The working title for the final chapter has long been *The Future of Life in the Hominoid Niche*. Yet, how can we advance a vision of "coexistence" among hominoid species when relationships within our own species are so frequently laced with structural racism, violence and inequality? What hope is there for nonhuman primates – what protections from humankind's planetary alterations can be ensured – while systemic forces within our own species can be so forcibly *dehumanising*? The one lesson anthropology teaches us with certainty is that alternative ways of living are possible; and therefore a re-arrangement of relations is always an option. An engagement with a dialectical primatology offers a possibly productive pathway towards such an option. Ape extinctions are ultimately inevitable; however, our immediate actions can prolong their existence. It is within our collective capacity, and indeed our responsibility, to do so.

Notes

1 For complete details see: Great apes, COVID-19 and the SARS CoV-2: Joint Statement of IUCN SSC Wildlife Health Specialist Group and the Primate Specialist Group, Section on Great Apes released on 15 March 2020. While it is still unknown if great apes are susceptible to the SARS CoV-2 virus, previous evidence exists with respect to the risk posed by infectious pathogens (e.g., Köndgen et al. 2008; Travis et al. 2008; Gillespie & Leendertz 2020).

2 Russell Mittermeier and Matthew Richardson begin their co-authored foreword to Colin Groves' (2008) book *Extended Family* with the following exchange:

> A few years ago, during a visit to Adelaide [ostensibly to attend the XVIII Congress of the International Primatological Society held in 2001], a reporter asked Jane Goodall what it felt like to be the world's foremost primatologist. "Oh no," she quickly replied, "You're mistaken. The world's foremost primatologist is Colin Groves, and he lives right here in Australia."

3 See Walker (1990), Smith (2013), Anderson et al. (2014) and Mutu et al. (2017) for a comprehensive assessment of the racialised structures within New Zealand society by leading Māori scholars and authors.

4 On 10 April 2019, less than a month after the murderous attack which left 51 people dead, the New Zealand Parliament passed (with a vote of 119–1) the Arms (Prohibited Firearms, Magazines and Parts) Amendment Bill banning most semi-automatic firearms, assault rifles and high-capacity magazines.

5 Also remember Ahmaud Arbery, Michael Brown, Eric Garner, Oscar Grant, Freddie Gray, Atatiana Jefferson, Trayvon Martin, Emmett Till and countless other lives lost to racialised structural violence.

6 In considering a core facet of our discipline – our evolutionary relationship to the apes – Jonathan Marks (2012: 97) asks the following, critical question: "Was winning the rhetorical battle against the creationists so crucial that we could afford to sacrifice the non-original sin of racism at our birth?" As Marks notes, even respected leaders in the field of evolution, human genetics and physical anthropology (e.g., Ernst Haeckel,

Charles Davenport) lacked a fundamental understanding of Darwinian tenets (the importance of both descent *and* modification), as well as the cultural and moral aspects of knowledge production, especially with respect to human evolution. Marks (2012: 100) observes that: "In particular, the first-generation German Darwinians managed to see continuity – with Africans intermediate between Europeans and apes – where in fact no continuity existed".

References

Anderson A, Binney J, Harris A. 2014. *Tangata Whenua: An Illustrated History.* Wellington: Bridget Williams Books.

Behie AM, Oxenham MF. 2015. *Taxonomic Tapestries: The Threads of Evolutionary, Behavioural and Conservation Research.* Canberra: Australian National University Press.

Beston H. 1928. *The Outermost House.* Reprint, with introduction by Robert Finch (1988). New York, NY: Henry Holt.

Fuentes A. 2011. Being human and doing primatology: National, socioeconomic, and ethnic influences on primatological practice. *American Journal of Primatology* 73(3): 233–237.

Gillespie TR, Leendertz FH. 2020. COVID-19: Protect great apes during human pandemics. *Nature* 579(7800): 497–498.

Groves C. 2008. *Extended Family: Long Lost Cousins. A Personal Look at the History of Primatology.* Arlington, VA: Conservation International.

Haraway D. 1989. *Primate Visions: Gender, Race, and Nature in the World of Modern Science.* London: Routledge.

Jost Robinson CA, Remis MJ. 2018. Engaging holism: Exploring multispecies approaches in ethnoprimatology. *International Journal of Primatology* 39(5): 776–796.

Köndgen S, Kuhl H, Goran P, Walsh P, Schenk S, Ernst N, Blek R, Formenty P, Maltz-Rensing K, Schwieger B, Junglen S, Ellerbrok H, Nitsche A, Briese T, Lipkin W, Pauli G, Boesch C, Leendertz F. 2008. Pandemic human viruses cause decline in endangered great apes. *Current Biology* 18: 1–5.

Lappan S, Malaivijitnond S, Radhakrishna S, Riley EP, Ruppert N. 2020. The human–primate interface in the New Normal: Challenges and opportunities for primatologists in the COVID-19 era and beyond. *American Journal of Primatology* 82: e23176.

Latour B. 2000. A well-articulated primatology: Reflections of a fellow-traveller. In Strum SC, Fedigan LM. (Eds.), *Primate Encounters: Models of Science, Gender, and Society* (pp. 358–381). Chicago, IL: University of Chicago Press.

Malone N, Wade AH, Fuentes A, Riley EP, Remis MJ, Jost Robinson C. 2014. Ethnoprimatology: Critical interdisciplinarity and multispecies approaches in anthropology. *Critique of Anthropology* 38: 8–29.

Marks J. 2012. Why be against Darwin? Creationism, racism, and the roots of anthropology. *American Journal of Physical Anthropology* 149(S55): 95–104.

Mutu M, Pōpata L, Williams T, Herbert-Graves A, Rēnata R, Cooze J, Pineaha Z, Thomas T, Kingi-Waiaua T. 2017. *Ngāti Kahu: Portrait of a Sovereign Nation: History, Traditions and Tiriti o Waitangi Claims: Kia Pūmau Tonu te Mana Motuhake o ngā Hapū o Ngāti Kahu: Ngāti Kahu Deed of Partial Settlement.* Wellington: Huia Publishers.

Riley EP. 2020. *The Promise of Contemporary Primatology*. New York, NY: Routledge.

Smith LT. 2013. *Decolonizing Methodologies: Research and Indigenous Peoples*, 2nd ed. London: Zed Books Ltd.

Travis D, Lonsdorf EV, Mlengeya T, Raphael J. 2008. A science-based approach to managing disease risks for ape conservation. *American Journal of Primatology* 70: 766–777.

Walker RJ. 1990. *Struggle Without End: Ka Whawhai Tonu Matou*. Auckland: Penguin.

Index

adaptive complementarity *see* dialectal approach
Addison, C. 140
Afropithecus 30
agile gibbon *10*, **35**
Anak Krakatau, Indonesia 111–112
Angola 43
animal ethics committees (AECs) 143, 145–146
animal rights 129
animals: anthropomorphism 8; dualisms 102; empathy 171; myths 106, 118; models 141; prey 46; symbols 105; value of 146–147; *see also* Zoological gardens
Anthropocene: geological epoch 6, 15, 65, 82, 103, 119, 130, 136, 161, 163, 173; conceptual debates 161, 172
anthropocentrism 170–171
anthropogenic: alterations 16, 73–75, 77, 82–83, 97, 99–105, 109–113, 130–132, 136, 141–142, 160, 165, 169, 172; displaced apes 2, 13, 110–111, 130, 132, 134, 136–137, 161, 166; primate hunting 73, 81–83, 110, 112, 116–117, 131, 137–139, 142, 144, 164, 168; primate trade 2, 9–10, 12, 15, 110–111, 116, 130, 132, 139, 143, 164; *see also* coexistence; grey sites
anthropology: primatology as 3, 65, 103–104, 130–131, 148; race 183–184; sociocultural 113, 118, 136
Aotearoa *see* New Zealand
Arabian Peninsula 30
Auckland Zoo, New Zealand 13, 132, 144–145, 171
Australia 134, 143, 168, 182

BaAka people, Central African Republic 139

Bangladesh 38
banteng 96, 105, 109, 112–113, 118
biomedical research *see* ethics
Bos javanicus see banteng
Batang Toru, Indonesia 41
biocultural evolution 163
Brazil 98
Bossou, Guinea 46
boundaries: conceptual 26–27, 104, 121n7, 133, 166; scholarly 8; social networks 72; spatial 101, 112–113, *117*, 137; temporal 104; *see also* transvalued species concept
Bukit Lawang, Sumatra *42*
Bali, Indonesia 9, 15, 98
behavioural ecology 11, 35, 38, 66–68, 70, 81, 117, 165
bonobos 2, 12, 26, 33, **40**, 45–47, 75; social structure compared to chimpanzees 78–80; *see also Pan* spp.
behavioural flexibility 15, 46–49, 66–83, 144, 160–162, 164; *see also* constrained totipotentiality; dialectal approach

Cagar Alam Leuweung Sancang (CALS), Indonesia 1, 96, 106–111, 117–118, 133; comparison to Ujung Kulon 114–116; silvery gibbon niche construction 76–78; *see also Hylobates moloch*
Cameroon 43, 141–144, 169, 183
captivity: aggression 79; ambassadors 145, 171; as a categorical state 130, 132, 166; behavioural enrichment 9; bonobos 79; breeding 134–135; chimpanzees 9; conservation 135; dichotomy 4; ethics 2, 15, 130, 166; *naturalcultural* space 166; orangutans 42, 81, 132, 145; silvery gibbons 133

catarrhine 28–29, 98, 163
Central African Republic 43, 138–139
Cercopithecoidea 28–29, 98, 163; *see also* monkeys
chimpanzee 2, 26, 33, **40**, 45–47, 72, 147, 172: Chimpanzee and Human Communication Institute (CHCI) 9; Ebola vaccination 140–141; Nigeria-Cameroon chimpanzee 13, 141–144, 169–170; social structure compared to bonobos 78–80
China: fossil apes 30, 33; gibbons 26, 38; orangutans 82
Ciwidey, West Java 133, 137
coexistence: ethnoprimatology 64, 103–104, 119, 130–140, 148, 166, 183; extinction crisis 147, 161, 173; human-ape interface 8, 13, 15–16, 71–72, 103–105, 118–120, 130–136, 141–144, 166–169, 182; hominoid 8, 14–16, 27, 147–149, 161, 164–165, 167–172, 183–184; natureculture 102–120; resilience 16, 27, 64–65, 75, 80–84, 117–118, 146, 163; *see also* conservation; ethics
Congo Basin 12, 46
conservation: conservationists 3–4, 7–8, 13, 15, 41, 50, 113, 118, 142, 148, 161, 168, 182; conservation strategies 7–8, 16, 117, 167–170; interventions 130, 133–135, 144, 162, 164, 167; intrinsic value 119, 138–139, 168; primate extinction risk 7, 14, 16, 27, 38, 50, 83, 98–99, 116, 145–148, 161, 165, 171–173, 184; protectionist approach 135, 167–168; research approaches 2, 4, 12–16, 50, 77, 81, 84, 96–97, 130, 135–143, 147–149, 162; *see also* coexistence
constrained totipotentiality *see* dialectical approach
constructed compatibility *see* dialectical approach
critical political ethology 65, 73–83, 164
Cross River gorilla 13, 43, 141–143, 169
cultural mapping 108, 122n15, 137

Darwin, C. 5, 66
Darwinian evolution 68, 185; *see also* population-level processes
deforestation 81, 99–101, 109, 114, 116, 166

Democratic Republic of the Congo (DRC) 2, 43, 79, 98
dialectical approach: as a primatologist 8–12, 83–84, 130–136, 139–148, 161–162; constrained totipotentiality 4, 18n5, 29, 33, 49, 65, 68–69, 98–99, 119, 162, 173; constructed compatibility 4, 28, 48–49, 66–73, 75, 81–82, 119, 162–163; dichotomies 4, 7, 70, 119, 132, 148, 167, 171; emergence 4, 7, 14–15, 34, 49–50, 68–78, 119, 162–164; habits of thought 3–4, 119; philosophical positions 4–7; principles of 4, 7, 27, 65, 119, 162; *see also* reductionism
Dipterocarpaceae 1, 41, 108
Dryopithecus spp. 31
Dzanga Sangha, Central African Republic 138–139

Economic dynamics 130, 148; Cameroon 142; Java 100–102, 107, 117–119; *see also* transvalued species concept
Ekembo 29
emergence *see* dialectical approach
Equatorial Guinea 43
ethics: biomedical 130; Ebola vaccine 140–141; field research 15, 130, 132, 140, 143–144; human participants 135, 143–146; managed setting 144–147; *see also* animal ethics committees
ethnographic insights: 12, 15, 84, 103, 108, 114–116, 131–132, 135- 140, 144–147, 165-7, 169; *see also* ethics
ethnoprimatology *see* coexistence
Eurasia 29, 32
Europe 30–32
extended evolutionary synthesis (EES) 14, 69–73, 75, 80, 83, 164; *see also* dialectical approach
extinction filtering 82–83

Fongoli, Senegal 46
forelimb morphology 28, 30–31, 34, 163
Fuentes, Agustín 9, 12, 36, 42, 70, 148, 162

Gabon 43
galagos 46
Ghana 168
ghost lineage 32–33
gibbon *see* Hylobatidae
Gigantopithecus 32

Gorilla spp. 2, 13, 26, 33–34, 40, 43–44, 168–169; field research 139, 141–143; zoos 129, 134, 146; *see also* Cross River gorilla
Gould, S.J. 4–5, 70, 161
Graecopithecus see Ouranopithecus
gray sites 15, 130, 132–135
great apes *see* Hominidae
Greece 32
Gunung Honje ranges 112–115

habituation 46, 140, 167
Harambe 128, 146, 149n1
Hegel, G.W.F. 4, 6
Hegelian triad 83
Heliopithecus 30
Holocene 29, 98, 163
Hominidae 14, 32–34, 40–49
Homininae 14, 31–32, 34, 40, 43–48
Hominini **34, 40**, 47–48
hominoid grade shift 49
Hominoidea 3, 6, 14, 16, 26–50, 69, 71–73, 80, 98, 160–163, 167, 172–173
Homo spp. 26, **34**, 40, 47–48
Hoolock spp. **34, 35**, 37, 38
Human Participant Ethics Committees (HPECs) *see* ethics
human–ape interface *see* coexistence
Hylobates agilis 10, **35**
Hylobates lar **35**, 36–37
Hylobates moloch 11, **35**, 76–77, 96, 106–110, 114–117, 133–134, 165
Hylobates spp. **34**, 35–37
Hylobatidae: behavioural plasticity 7, 72, 75–78; Chinese culture 26; cognitive capacity 49; effects of seasonal environment 98; genera 35–40; Hylobatidae - Hominidae divergence 32–33; illegal trade of 9–11; taxonomy 34–35

India 31, 38
infanticide 77, 79, 85n9
Institutional Review Boards (IRBs) *see* ethics
International Union for the Conservation of Nature (IUCN) 106, 137, 181
island biogeography theory 98–99

Java, Indonesia: biogeography 97–99; bird markets 9–10; political economy and ecology 11, 99–102, 169; primates 15, *36*, 40, 82, 84, 112, 116, 133–139, 165–166; scared forests 1–2, 12, 96, 104–107, 110–111, 117–119; tsunamis 111–112, 166; *see also* Cagar Alam Leuweung Sancang, natureculture, Ujung Kulon National Park
Javan lutung 104, 106
Javan Primates Rehabilitation Centre (JPRC), Indonesia 110, 133, 137–138
Javan slow loris 106
Javan surili 106, 116
Jost Robinson, Carolyn 138–139

Kenya 29–30, 32
Kenyapithecus 30
keramat *see* sacred sites
Ketambe Research Station, Gunung Leuser National Park, Sumatra 36, 80
Khao Yai National Park, Thailand 36
Krakatau Islands Nature Reserve, Indonesia 112
Kroeber, Alfred 17

Levins, Richard 3–5, 70, 83, 161
Lewontin, Richard 3–5, 70, 83, 161
life history 26, 34, 98, 160
Lomako Forest, Democratic Republic of the Congo 12, 79
Lore Lindu National Park, Sulawesi, Indonesia 138

Macaca fascicularis fascicularis 106
Macaca tonkeana 138
macaque 13, 106, 138
Madagascar 98
Malay Peninsula 37
managed settings 15, 130, 132–135, 144, 166, 168
Martu people, Western Australia 168
Marx, Karl 4–5, 68
Miocene 27–33, 47, 72–73, 162–163
Modern Synthesis 5, 66, 70
Mone-Oku Forest, Cameroon 142–143, 169
monkeys 12, 49, 98, 144, 163, 182; divergence 28–29; Javan monkeys 106; monkey parks 132
monogamy 39, 77, 85n4, 86n13
Morotopithecus 29
Mount Tilu Nature Reserve 133–134
multispecies approaches 117, 130, 136–139, 147, 170–171, 173
Myanmar 38

Nacholapithecus 30
natureculture *see* coexistence
New Zealand 132, 140, 143, 145,
    181–183
niche construction theory (NCT) 48,
    71–78, 80, 164
Nigeria 43, 168
Nomascus spp. **34**, **35**, 37, 38–40
Nycticebus javanicus 106
natural selection 5, 14–15, 66–67, 71–72,
    77, 81

Oates, John 119, 134–135, 168–169
Oligocene 28, 163
One Health (OH) 140
Optimal behavioural strategies 49, 68–71
orangutan: Bornean orangutan 40–41;
    conservation 148; in zoos 9, 13, 132,
    134, 144–145, 171; Java 99; relationship
    to *Sivapithecus* 31–32; social and
    ecological variation 40–42; social
    flexibility 75, 80–83, 164; Sumatran
    orangutan 40–41, *42*, 134; Tapanuli
    orangutan 40–41; taxonomy 14, 26,
    33–34
Ouranopithecus 32

Pakistan 31
Paleotropical monkeys *see*
    Cercopithecoidea
Palmer, Alexandra (Ally) 132–133, 140,
    144–145, 171
Pan paniscus see bonobo
Pan spp. *31*, **34**, 40, 45–47, 78–80, 83, 164
Pan troglodytes see chimpanzee
Panini *see Pan* spp.
Panthera pardus melas 105, 116
Panthera tigris sondaica 99, 103–106,
    109–110, 118
Pierolapithecus spp. 31
Pleistocene 28–29, 33, 82, 98–99, 163
Pliocene 28–29, 163
political ecology 65, 84n3, 164–165
political economy 65, 84n3, 99–102,
    142–143, 165
Pongo spp. *see* orangutan
population-level processes 29, 36, 39, 49,
    68, 71, 77
Presbytis comata 106, 116
primate cognition 26, 34, 40, 48–49,
    73–74, 160
primatology: academic lineages 12–13;
    cultural endeavour 3, 27, 172; field

primatology 3, 130, 140, 149n5, 167;
    future of 14, 97, 119–120, 140, 161,
    173, 181–182, 184; traditions 85n7,
    182; *see also* anthropology; coexistence;
    dialectical approach; ethics; extended
    evolutionary synthesis
Proconsul spp. 29–32

quadruped 29, 31–32

reductionism 4–5, 66–67, 70, 83
reintroduction 15, 77, 79, 81, 110,
    132–139, 166
Remis, Melissa 138–139
rhinoceros 103, 112–113, 118
Riley, Erin P. 4, 131, 138, 148
"Robinsonades" 68
Rwanda 43
rehabilitation 13, 15, 77, 110, 130,
    132–138, 166

Sabah, Malaysia 41
sacred sites (Java) 96, 107, 109–110,
    117
Samburupithecus 32
Sarawak, Malaysia 41
Selby, Megan 106, 164
self-domestication hypothesis (SDH)
    78–80
siamang 26, 33, **35**, 37–38, 39
silvery gibbon *11*, **35**, 76–77, 96,
    106–110, 114–117, 133–134,
    165
Sivapithecus spp. 31–32
social organisation, primate 36, 38–39, 42,
    44, 48, 66, 68, 75, 81, 85n4
social relationships, primate: cooperative
    behaviour 7, 70, **71**; evolved emphasis
    34, 49, 73, 160; social bonds 7, 19n11,
    35–39, 71, 133–134; social networks
    7, 37, 44, 48, 72; social niche 72–73,
    78–80; social plasticity 44, 72–74;
    traditions 42, 47, 49, 78–80; *see also*
    social organisation; socioecological
    models
socioecological models 12, 27, 65–83
    behavioural adaptations: 66, 76;
    feedback loops: 68, 72–73, 77, 80, 118;
    see also dialectical approach; extended
    evolutionary synthesis
Spain 31
spiritual pilgrims 1, 96, 107, 109–110,
    114–117, 133

Sponsel, Leslie 131
standard evolutionary theory (SET) 14,
    65–71, 83
Suaq Balimbing, Sumatra 42, 80
Sumatra, Indonesia 11, 13, 36–37,
    40–42, 81, 98–99, 111, 134,
    166
Sunda Shelf 40, 98
Sundanese people, West Java 1, 102, 104,
    106–107, 116, 169
Symphalangus syndactylus 26, 33, **35**,
    37–38, 39

Taylor, Bron 146
Tonkean macaques 138
Trachypithecus auratus 104, 106
Tulp, Nicolaes 147
Turkey 30
Tyson, Edward 147

Uganda 29, 43, *45*
Ujung Kulon National Park, Indonesia
    105, 106, 110–118
University Research Ethics Committees
    (URECs) *see* ethics

von Liebig, Justus 5

Wade, Alison 13, 136–137, 141–144, 169,
    183
Wallace, Alfred Russel 5, 66, 82, 103
Way Canguk Research Station, Sumatra
    37–38
Wessing, Robert 104–105
Wheately, Bruce 131
White, Frances 2, 12–13, 42, 70
white-handed gibbon **35**, 36–37
Wildlife trade *see* anthropogenic

Yuanmoupithecus 33
Yunnan, China 33

zooanthroponosis 164, 167
Zoological gardens: anthropocentrism
    170–172; Auckland 13, 132, 144–145,
    171; Brookfield 9; captive breeding
    134–135; Cincinnati 129, 146–147;
    Denver 8; ethics 129–130, 144–147;
    role in conservation 15, 167; *see also*
    coexistence
zoonoses 6, 181

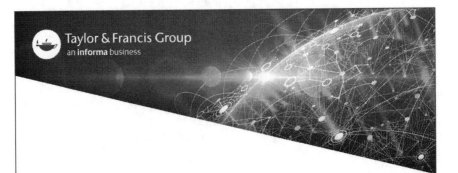

Taylor & Francis Group
an **informa** business

Taylor & Francis eBooks

www.taylorfrancis.com

A single destination for eBooks from Taylor & Francis
with increased functionality and an improved user
experience to meet the needs of our customers.

90,000+ eBooks of award-winning academic content in
Humanities, Social Science, Science, Technology, Engineering,
and Medical written by a global network of editors and authors.

TAYLOR & FRANCIS EBOOKS OFFERS:

A streamlined
experience for
our library
customers

A single point
of discovery
for all of our
eBook content

Improved
search and
discovery of
content at both
book and
chapter level

REQUEST A FREE TRIAL
support@taylorfrancis.com

Routledge
Taylor & Francis Group

CRC Press
Taylor & Francis Group